MARKING MODERN TIMES

Marking Modern Times

A HISTORY OF CLOCKS, WATCHES,
AND OTHER TIMEKEEPERS IN AMERICAN LIFE

Alexis McCrossen

THE UNIVERSITY OF CHICAGO PRESS | CHICAGO AND LONDON

The University of Chicago Press, Chicago 60637
The University of Chicago Press, Ltd., London
© 2013 by The University of Chicago
All rights reserved. Published 2013.
Paperback edition 2016
Printed in the United States of America

25 24 23 22 21 20 19 18 17 16   2 3 4 5 6

ISBN-13: 978-0-226-01486-9 (cloth)
ISBN-13: 978-0-226-37968-5 (paper)
ISBN-13: 978-0-226-01505-7 (e-book)
DOI: 10.7208/chicago/9780226015057.001.0001

Library of Congress Cataloging-in-Publication Data

McCrossen, Alexis.
    Marking modern times : a history of clocks, watches, and other
timekeepers in American life / Alexis McCrossen.
        pages : illustrations ; cm
    Includes bibliographical references and index.
    ISBN 978-0-226-01486-9 (cloth : alk. paper)
    ISBN 978-0-226-01505-7 (e-book)
    1. Clocks and watches—United States—History—19th century. 2. Clocks
and watches—United States—History—20th century.   I. Title.
TS543.U6M396 2013
681.1′1—dc23

2012045005

♾ This paper meets the requirements of ANSI/NISO Z39.48-1992
(Permanence of Paper).

To Adam and Annie

CONTENTS

ILLUSTRATIONS

TABLES

FIGURES

ACKNOWLEDGMENTS

Robert Devens, senior editor at the University of Chicago Press, started talking with me about *Marking Modern Times* years ago. Even when a baby, politics, and another book on another topic with another press distracted me, he kept up the dialogue. As I was finishing the manuscript, Robert continued to shape my thinking and to offer invaluable help, particularly with chapter titles. I extend my thanks to him and to his colleagues, especially manuscript editor Carlisle Rex-Waller, for their professional, courteous, efficient, and kind help bringing this book to press.

Archivists and librarians helped me excavate the building blocks for this book: I thank especially the staff at the American Antiquarian Society, Baker Library's Historical Collections at the Harvard Business School, Historic Northampton, Independence Hall National Historic Park's Archives, the Library Company of Philadelphia, the Library of Congress, the United States National Archives, the National Association of Watch and Clock Collectors Library, the New-York Historical Society, the New York Public Library, the Archives Center, Library, and Division of Technology at the Smithsonian Institution's National Museum of American History, and Southern Methodist University's Bridwell Library, Fondren Library, and DeGoyler Library.

Several organizations made it possible for me to do extensive research into the history of timekeeping. The National Endowment for the Humanities funded a year (1999–2000) during which I spent nearly every day in archives. It was during this year that I discovered some of the book's key protagonists: watch repairman David E. Hoxie, public clock manufacturers Seth Thomas and E. Howard, countless time ball officers, standard time's advocate, William F. Allen, and the inestimable Francis G. DuPont. Laurel Thatcher Ulrich and Lizabeth Cohen, directors of Harvard University's Charles Warren Center, provided me with an institutional home for part of my NEH fellowship year. Harvard Business School's Laura Linnard and Walter Friedman made that home all the more welcoming. Research grants from the Smithsonian Institution's Lemelson Center for the Study of Invention and Innovation, the Hagley Museum and Library, SMU's University Research Council, and SMU's Clements Department of History extended the amount of time I was able to spend in archives and at libraries and provided resources for the book's illustration.

I thank the editors of *Material History Review, Journal of Urban History, Winterthur Portfolio*, and *Common-place* for permission to draw on and rework my articles on timekeeping that they published. Parts of chapters 2, 4, and 5 are adapted from "Time Balls: Marking Modern Times in Urban America, 1877–1922," *Material History Review* 52 (2000): 4–15. Parts of chapter 3 draw on "The 'Very Delicate Construction' of Pocket Watches and Time Consciousness in the Nineteenth-Century United States," *Winterthur Portfolio* 44 (Spring 2010): 2–30. Parts of chapters 4 and 5 are adapted from "'Conventions of Simultaneity': Time Standards, Public Clocks, and Nationalism in American Cities and Towns, 1871–1905," *Journal of Urban History* 33 (January 2007): 217–53. Several chapters draw on an essay about the sound and look of time for which Cathy Kelly generously made room in the October 2012 issue of *Common-place*.

I delivered many papers based on the research for this book. I extend my thanks to my hosts and audiences at the Clinton Institute for American Studies in Dublin, Ireland; American University's Department of History; the University of New Mexico's Department of History; Pennsylvania State University's Civil War Era Center; the Independence National Historical Park in Philadelphia; the Dallas Area Social History Group; SMU's Department of History; the Smithsonian Institution's Lemelson Center; the Newberry Library's "Technology, Politics, and Culture" seminar; Harvard University's Charles Warren Center; the Society of Nineteenth-Century Americanists 2012 Annual Meeting; the American Historical Association's 2006 and 2001 Annual Meetings; the Vernacular Architecture Forum's Twenty-Fifth Annual Meeting; the American Studies Association's 2001 Annual Meeting; the Group for Early Modern Cultural Studies' 2000 Annual Meeting; and the International Society for the Study of Time's 2001 and 1998 Triennial Meetings.

At various stages of research and writing, friends and colleagues have commented on the whole or parts of this manuscript. I thank Tom Allen, Christopher Clark, David D. Hall, Walter Friedman, Amy Greenberg, Adam Herring, Katie Lofton, Michael O'Malley, Jarrett Rudy, Mark M. Smith, Carlene Stephens, Cheryl Wells, and Nick Yablon for reading and commenting on versions and sections of the manuscript. The book is considerably better owing to their quick eyes, interpretive acumen, and generosity. I also thank for suggestions and inspiration my colleagues in the History Department at SMU, especially past and present chairs Jim Breeden, Jim Hopkins, Dan Orlovsky, Sherry Smith, and Kathleen Wellman. Sharron Pierson and Mildred Pinkston, the History Department's staff, were helpful at every turn, whether with planning research trips, filing reports, making copies, or just listening.

Friends and colleagues who have contributed to this book's genesis over the years include Alison Alonso, Amy Amend, Adam Arenson, Tom Augst, Susie Bajari, Ian Bartky, Francesca and Del Beveridge, Pleun Bouricius, Rob Bruegmann, Jill Callahan, Gregg Cantrell, John Chavez, Guy Chet, Stephanie Cole, Ed Countryman, Maggie Dennis, Chris DeSantis, Juliana DiGiosia, Melissa Dowling, Derick Dreher, Amy Earls, Daniel Garza, David Goldfield, Kathy Grover, Kenneth Hamilton, Meg Jacobs, Richard John, Tom Knock, Russell Martin, Rob Maxwell, Jessica May, John Mears, Laura Milsk, Francesca Morgan, Walton Muyumba, Michele Nickerson, Farhad Niroomand, Julia Ott, Alison Parker, Griselda Perez, Raul Ramos, Hai Ren, Joan Rubin, Beth Savoldelli, Sarah Schneewind, Stephen Sennott, Laura Isabel Serna, Ling Shiao, Erin Smith, Kurt and Birgit Stache, Kirsten Sword, Roberto Tejada, Charissa Terranova, Liz Turner, Chris and Tara Twomey, Dan Wickberg, Michael Zakim, and Christina Zienkowsky. I am grateful for the interest they took in my work, and for their collegiality and friendship.

I would be remiss were I to pass over the opportunity to formally acknowledge the guidance and inspiration of a few people in particular. For nearly a quarter of a century, I have benefited from David D. Hall's mentorship, scholarship, and friendship. David's expertise in material culture, in the history of ideas, in New England, and in the nineteenth century has shaped mine for the better. When I was just beginning my research, I met the wonderful Carlene Stephens, the curator in charge of the clocks and watches at the Smithsonian Institution's National Museum of American History. I am thankful to her for her intellectual and personal generosity. A historian of nineteenth-century American literature, Tom Allen, has contributed in innumerable ways to this manuscript. He is a singularly gifted reader, as well as an ideal interlocutor. I am grateful for Crista DeLuzio's intellectual, professional, and personal support for the long time we've known each other; I look forward to many more happy years of collaboration in our various endeavors. David Doyle Jr. and I enjoyed a research trip to Northampton while staying at the delightful country home of Pleun Bouricius one summer long ago. David reads books the way he treats friends—with respect, loyalty, and affection. Amy Greenberg's friendship and wisdom over these many years has helped to sustain me; her prowess as a historian has inspired me; her critical insights have enriched me. My extended family—the Cammermeyers, the Fosters, the Herrings, the McCrossens, my mother Macon McCrossen, my brother-in-law Nate Orr, and my sister Tamara McCrossen-Orr—gave me places to stay, fed me plentiful meals, plied me with drinks, offered me companionship, drove me to many places, lent me their cars, and even provided me with desks at which I wrote and revised most of this book. I am deeply thankful.

In closing, I thank my daughter Annie Herring and my husband Adam Herring. The love each evinces for language, images, and ideas is infectious. At once I am grateful for all the time together and for all the time apart. Without the former I'd be bereft; without the latter this book would not have made it into print. Annie's fine company enlivened inevitable weekend detours to my office, the UPS store, and the library. She vetted the book and chapter titles. She built a potato clock. Adam made it possible for me to do the research, build the databases, read the secondary literature, gather the illustrations, make false and fresh starts at the writing, and revise again and again. Not only did he open up spaces of time for me, he consulted about images, helped me choose just the right words, and recognized my ideas as consequential. I am forever in Adam and Annie's debt and wish I could stop the hands of time from moving forward so we could live together as we are forever. To them I dedicate this book.

Unveiling the Jewelers' Clock

On a winter's day in 1928, a prosperous group of jewelers in Chicago unveiled a magnificent four-faced clock affixed to an ornate skyscraper known as the Jewelers Building.[1] The Elgin National Watch Company, headquartered in the building, sponsored the installation of the six-ton clock made of steel and bronze. The clock itself was an elaborate period piece: illuminated with electric lights, decorated in the Oriental motifs favored in twenties' design, and topped with a figure of Father Time holding a scythe and an hourglass. It brought the Jewelers Building, a Renaissance-revival confection wedged between the Chicago River and downtown, into spatial and temporal relation with several landmark buildings on the edge of Chicago's loop. The clock and bell tower of the grand courthouse where Lincoln's body lay in state in 1865 would have been just two city blocks distant were it not for the Great Fire of 1871. Not far away were Chicago's five railroad stations; some with towering clocks, others flanked by impressive post clocks. Only a handful of Chicago's massive commercial buildings featured exterior clocks; these few were typically suspended from elaborate brackets over sidewalks, as were the famous clocks hung from Marshall Field's department store in 1898 and 1907. A few hundred yards upriver from the Jewelers Building, a 1914 tower clock crowned a red brick warehouse belonging to Reid Murdoch and Company, a wholesale grocery firm. Less than a mile downriver, four clock dials spanning nearly twenty feet each adorned the 425-foot-high south tower (completed in 1921) of the gleaming Wrigley Building.

Tarp being pulled off a new clock, Chicago, 1928, *Chicago Daily News* photograph, DN-0084777, Chicago History Museum. Father Time is perched atop this spectacular four-faced clock whose unveiling came at the end of the public clock era.

The words "Elgin" and "Time" appeared in gilded capital letters above and below the clock face. They were reminders of the days when timekeeping was the purview and privilege of horologists, not of wholesalers or chewing-gum magnates. The jewelers who crowded beneath the Chicago clock when it debuted might have recalled the years when they belonged to the ranks of time specialists, those artisans, tinkerers, and scientists who reckoned the time and made clocks for its conveyance. Since the seventeenth century, they had monopolized the trade in small timekeepers, offering new, used, and re-furbished clocks and watches in a variety of casings. As artisans and tinkerers, they had developed specialized knowledge related to the workings of me-chanical timekeepers, which for centuries had been novelties, ornaments, and playthings. Jewelers also regulated, maintained, and repaired timepieces. It was jewelers too who sold nearly all of the American watches and watchcases manufactured in the large, highly capitalized factories that went into busi-ness in the northeastern and midwestern United States during the 1860s and 1870s. Among these enterprises was Elgin, whose wildly productive pocket-watch factory was along the Fox River, forty miles north of Chicago in the model factory town that bore the company's name. Its marketing included newspaper advertisements, advertising booklets, a magazine, and even a mili-tary marching band, but Elgin did not sell directly to consumers. Instead, jewelers sold Elgin's watches.

For centuries, jewelers, watchmakers, and their brethren clockmakers had also fulfilled another important civic duty: they determined the local time using astronomical instruments and charts. In the 1840s and 1850s, self-styled astronomers encroached with the construction of permanent observa-tories and time services that made use of the telegraph. In the US after the 1880s, when a national standard time calibrated to Greenwich meridian time replaced local time, jewelers' status as local timekeeping authorities further waned. Most of their clocks could not compete with the imposing time-pieces found on government and commercial buildings and connected by telegraph to astronomical observatories. As the US government, corporate entities, and various scientific bodies amassed authority for the time, jewelers' temporal authority came more from the sale and repair of timepieces than from the public service of determining and distributing the local time. Their status as public servants had thus been diminished, but with the magnificent Jewelers Building clock, they could at least gesture to that lengthy epoch of influence.

The unveiling of the jewelers' clock happened at the end of what I call the "public clock era," the subject of this book. During the previous century or so, commercial and civic institutions sponsored the installation and regula-tion of thousands of public clocks. In the attempt to coordinate these clocks, they muddled through different processes of standardizing time and experimented with procedures and technologies for distributing the time. All the while men, women, and children acquired pocket watches of various makes, signatures, and materials. They decorated their homes with clocks; included clocks and watches in their poetry, sermons, stories, and songs; and enlivened their visual culture with representations of timekeepers. Like the clock on the Jewelers Building, the casing, cabinetry, and design of these timepieces deployed symbols, quite often nationalist, that ascribed transcendent significance to mechanical time.

The fanciful Jewelers Building clock was set going near the end of a long period of nation building and imperial activity around the world. Clocks, as adjuncts to armies, bureaucrats, and other agents of state building, played a role in the consolidation and expansion of both nations and empires. Time-reckoning technologies and practices are an age-old way to assert political power and mastery.[2] In the United States, governing authorities were initially slow to assert authority over the measurement and distribution of time, but after the Civil War their pace hastened. Elsewhere around the world, central authorities have issued edicts about the time in the effort to consolidate power. Chairman Mao placed all of China, which geographically covers eight time zones, under Beijing time in 1949. More recently, in 2011, Dmitry Medvedev made daylight saving permanent in Russia and eliminated two of the enormous country's eleven time zones. Jurisdiction over the time was one of the few powers President Medvedev exercised while in office, much to the frustration of Russians, who lived in darkness until nine or ten in the morning during that 2011–12 winter.[3]

Within this context where timekeepers were at once political instruments, social tools, and cultural symbols, public clocks habituated Americans to "clock time," which encompasses the practices and material culture associated with mechanical timekeepers.[4] Local knowledge gradually accumulated about where public clocks could be found, who owned and maintained them, how well they ran, and which events were subject to the time they meted out.[5] So habitual, so small, and so repetitive as to leave few traces in historical sources, these practices—looking to clocks for the time; comparing clocks with the sun, the stars, and each other so as to find deviance and constancy; measuring duration with clocks; assigning clock time to events; using clocks to coordinate social action—constitute the core of clock time. Clock time was rooted in experiences like checking a pocket watch after hearing a known bell sound the hours, or looking up at a reputable clock to ascertain the exact minute, or finding a good vantage to see a time ball drop. More than any other device or circumstance, public clocks made clock time palpable, possible, and even desirable. They culturally and phenomenologically tied mechanical time to particular narratives of experience, spaces, and institutions. Public clocks presented clock time as unfolding, site specific, and institutional.[6] They distilled pressures and ambitions that inspired people to acquire timepieces of their own. The far-reaching implications of this profoundly important system of social regulation, whose emergence depended on public clocks, pocket watches, and standard time, helped to define modern times.

Public clocks like the one hanging from the Jewelers Building have a lengthy history rooted in the human impulse to mark time in one way or another and to accrue power through control over the time. They date to the sundials, clepsydras, and incense clocks of antiquity and to the earliest mechanical clocks of the twelfth century.[7] After the thirteenth century, clocks and bells indicating the hours spread through villages, small towns, and urban centers in Western Europe. The parameters within which the earliest clocks functioned were broad: a hand moved either direction (rather than only "clockwise"), and dials indicated three, four, sixteen, or twenty-four hours, in addition to the now conventional twelve hours. By the end of the fifteenth century, hands moved clockwise on dials with twelve equal hours.[8] In early modern Europe, the public clock was "an indicator of modernity," like the period's other technological and social innovations, mills and schools. Public clocks dotted the public spaces of capital cities like Paris, London, and Berlin; port cities like Bristol, England; and colonial cities like Quito, Tlaxcala (Mexico), New York, Philadelphia, and Boston.[9] By the middle of the eighteenth century, clocks constituted the spatial and symbolic center of urban life in Western Europe, Spanish North and South America, and British North America.

The audibility and visibility of hour's bells and clock faces made them public, regardless of ownership, provenance, or location. In the middle of the fourteenth century, the term "public clock" began to be used to refer to a clock found on the interior or exterior of a building to which a public had access.[10] Traditionally the technologies of keeping time and telling time were separate; a bell might have rung the hours, but it did not compute the time. Clocks brought the two technologies together: they kept and told the time. In this book, I use the category "public clocks" to describe several distinct mechanisms, all of which told the time, only some of which kept it. My inquiry into American public clocks includes bells, mechanical clocks, and time balls.

Many obstacles stand in the way of sketching out the processes through which public clocks saturated American places. The only book-length study of public timepieces in the United States, Frederick Shelley's useful survey, *Early American Tower Clocks*, terminates its investigation in 1870. Establishing a complete census of timepiece installation in the United States is a tantalizing goal, so much so that some of the twenty thousand individual members of the National Association of Watch and Clock Collectors (NAWCC) have provided counts of public clocks in their local areas. Their extraordinary efforts remain incomplete and provisional.[11]

Fortunately, abundant quantitative and qualitative evidence can be found in the account books of two companies, Boston's E. Howard Clock Company and Connecticut's Seth Thomas Company. These two manufacturers held a duopoly over the manufacture and installation of public clocks between the 1870s and the 1930s, the height of the public clock era in the United States and indeed throughout the world. E. Howard and Seth Thomas supplied the clockworks, dials, faces, and striking mechanisms for chimes and bells for an estimated 15,000 public clocks erected in American towns and cities between 1871 and 1911.[12] In comparison, the output of other clockmakers was small, but still notable. Mathias Schwalbach, for instance, installed 55 public clocks between 1882 and 1915, mostly in Wisconsin churches. Nels Johnson, a self-taught clockmaker from Michigan, built 52 tower clocks between 1865 and 1912.[13] E. Howard and Seth Thomas provided all sorts of public clocks, but they specialized in the most prominent type: tower clocks.

So numerous were the clocks inside buildings during the public clock era that it is impossible to provide an estimate of their incidence. The democratization of governance and commerce over the course of the nineteenth century opened to the public innumerable interior spaces where clocks might be audible or visible. Manufacturing mechanisms that could run without heavy weights and long pendulums required extraordinary skill. So small clocks equipped with watch movements and tall-cased clocks with heavy weights and pendulums predominated: both were expensive and cumbersome. But innovations in the design and manufacture of clocks allowed their size to vary without affecting their accuracy. Furthermore, the development of electric circuits and batteries liberated clocks from cabinets altogether, meaning that they could be found almost anywhere inside a building.[14] Without a doubt, during the public clock era the vast majority of public clocks were found in interiors. As catalogs from the period demonstrate, thousands of designs proliferated, and hundreds of suppliers were in business. Interior clocks included tall-case clocks, gallery clocks, wall clocks of various sizes, and clocks in cabinets small enough to be placed atop furnishings. They could be found behind counters, affixed to or embedded in walls, suspended from ceilings, set above bank vault doors, sitting on desks and shelves. These clocks reinforced the dominion of public time extended by tower, street, turret, façade, and window clocks.

By the end of the nineteenth century, public clocks in the United States

Table 1. Types of clocks ordered from Seth Thomas and E. Howard, 1871–99, 1902–11

	Striking	Nonstriking	Total	Percentage of all orders
Tower	957	657	1614	52
Interior clock	28	842	870	28
Post	1	251	252	8
Façade	82	108	190	6
Not known	—	110	110	4
Clock system	2	43	45	1
Astronomical	—	23	23	0.5
Other	12	2	14	0.5
Totals	1,082	2,036	3,118	100

Source: Based on the account books of the Seth Thomas Clock Company, held by the National Association of Watch and Clock Collectors and the account books of the E. Howard Clock Company, held by the Archives Center, National Museum of American History, Smithsonian Institution. $N = 3{,}118$ orders.

Note: Table 1 draws on 1,994 of the 8,149 orders in the extant account books of the E. Howard Company covering the years 1871–1905 and on all 1,124 orders in the extant account books of Seth Thomas covering the years 1877–93 and 1903–11.

reached the peak of their practical and symbolic importance. In 1883, the nation divided into time zones wherein a national standard of time coordinated clocks both public and private. Thereafter the rate with which public and private entities installed clocks in American public spaces became ferocious. By 1900, clock faces could be found towering over urban neighborhoods, buttoning streets together, peeking out from lobbies, gracing entryways, gleaming in store windows, hiding under awnings. Frequently their works ran comprehensive time systems that synchronized the hands on dozens of dials hung within architectural spaces both grand and utilitarian, as well as on exterior surfaces, particularly towers. New and large bells, as well as automatic systems to trigger their chimes at appointed and regular hours, allowed the time to seep into places where clocks could not be seen. Additionally, small fortunes were spent on the construction and operation of time balls, a now obsolete time-disseminating technology. These timepieces aurally and visually saturated public spaces not only with indications of the time, but with messages about the cosmic and national importance of time itself.

As with the trajectories of many eras, the most grandiose expressions came near the end of the public clock era. The near mania for public timepieces in the United States coincided with a vogue for towers, monuments, and monumental forms.[15] In addition to installing thousands of workaday clocks in public spaces at home and abroad, civic and commercial institutions sponsored so-called monster clocks—the largest, most dramatic, and farthest-reaching timepieces of the public clock era. During the late decades of the nineteenth century and the opening ones of the twentieth, these monster clocks with gigantic dials and heavy hands hanging from skyscrapers captivated American attention. However, even as they were going up along with time balls and enormous bells, their days were nearly over. In the same years, wireless communications introduced what one historian has called "an invisible grid of time signals" that bypassed public clocks altogether.[16] Authoritative time signals and personal timekeepers became clock time's key devices. The new time regime—itself a product of scientific methods, widespread agreements distilled into maps showing time zones, and practices privileging clocks as arbiters of the time—no longer required monumental architecture, high-pitched science, loud bells, spectacular displays, small ornamental devices, or ideological sleights of hand. No amount of wizardry could shroud clock time's artifice; such wizardry did, however, make clock time a fundamental feature of modernity.

By the 1920s and 1930s, the end of the public clock era was at hand. As people moved through cities in electric streetcars, subways, commuter trains, and automobiles, they did not see or hear as many public clocks as they had as pedestrians or passengers on slow-moving horse-drawn vehicles. Skyscrapers and massive office blocks provided fewer opportunities to install visible clocks or audible bells. At or near eye-level, electric lights, billboards, and neon signs eclipsed most public clocks. The size and scale of interiors also dwarfed clocks, especially in commercial spaces like department stores, whose managers had given them over to mirrors, lights, and decorative schemes meant to stimulate the imagination. Although some clocks and clock towers were taken down during the public works projects of the 1930s and the scrap metal drives of the 1940s, many still remained at midcentury. Most of these were later demolished as part of the urban renewal efforts of the 1960s and 1970s. Today, the few remaining public clocks belong to institutions that trade on the aura of longevity—churches, colleges, and universities, and baseball teams.[17]

The near ubiquitous personal ownership of timekeepers and the ease of disseminating the correct time further rendered consultation of public clocks incidental. Wristwatches in particular allowed quick and easy access to the time. Heeding "the impulse to *wear* time," soldiers in the Anglo-Boer War (1899–1902) and the First World War (1914–18) were among the first to strap watches to their arms.[18] Within a few years, civilians started to wear wristwatches. In this new world of immediate access, men and women needed to know the time to the minute, to anticipate the arrival of particular moments, to marshal their time so that at the appointed hour they could be at their office desk, in the church pew, on the railroad platform, or seated in the movie theater. What is more, telephones and radios simplified the dissemination of time; individuals wishing to set their timekeepers could dial up time services or wait for the radio to announce the time. Today acquisition of the correct time requires only a glance at a cell phone or computer screen. Together changes in public space, new methods of precision time telling, and rising expectations for temporal coordination contributed to the demise of public clocks. But not to that of clock time.

With Father Time atop their clock, Chicago's jewelers drew attention to the cosmic significance of clocks and of clock time. The nineteenth century's rising density of timepieces in urban and rural settings reflected and reinforced a shift in authority for time, from nature—God, the sun, and the stars—to clocks, their owners, and time specialists. Across cultures and centuries, communities have devised ways to account for and measure the passage of time. Astronomical observations provided the basis for the divisions of time into endlessly repeating equal parts: today these units are years, weeks, days, hours, minutes, and seconds (months are the only contemporary unit of time whose duration varies). Calendars were fashioned to record years, months, weeks, and days; clocks to show smaller measures of time like hours, minutes, and seconds.[19] These units are time standards of a particular sort.

A time standard is not specific to clocks or industrialized societies. For centuries, the most common temporal points of reference were natural indicators: the size of the moon, the state of the tides, the height of the sun, the ripening of wheat, the activity of animals. When mechanical timekeepers were introduced to communities, they worked in concert with nature, usually indicating solar time with the hour hand, a service especially useful on cloudy days and at night. Their time was site specific. Even so, their hands did not exactly represent the sun's position, because from day to day and season to season the sun's place and duration in the sky changes. Mechanical clocks, made to keep track of uniform measurements (hours and minutes), cannot track variable measures. Here is where machine diverges from nature. Clocks and clock time divide the day into two sets of twelve equal hours, one set ante meridian, the other post meridian. Meridian here refers to the highest place the sun reaches in the sky—noon—but also to the longitudinal measurement situating a locale on the earth and thus in time.[20]

In order to achieve temporal regularity and uniformity in which each hour and each day is the same length, the clock must diverge from the sun. It is possible for a clock to keep apparent solar time, as did Yale College's in the 1820s, but mean solar time requires fewer adjustments to the clockworks than apparent solar time. Averaging sun time across the year results in "mean time," also known as "mean solar time." What is called "an equation of time" represents the difference in time between a clock following mean time and the sun; presented by almanacs in tables called "ephemerides," this equation provided a standard against which a clock could be set with reference to the sun alone. The sun will reach noon before or after the clock all but four times in a year.[21] Engraved on an American sundial dating to 1800, for instance, was an "Equation Table for Regulating Clocks" with the notations "Sun Slow" and "Sun Fast" arrayed around the dial. New York's local time provided the standard; the city's name was engraved above the twelve mark, set in temporal relation with a list of world cities that included Moscow, Constantinople, Venice, Madrid, New Orleans, Bombay, and Jerusalem. The warning "Noli confidere noctrem"—do not trust the night—was emblazoned across the top.

Each locale has a different "true time." A clock set to mean solar time at a particular meridian is said to be keeping "true time," which in the nineteenth century was used synonymously with the term "local time."[22] True time is important, for it helps geographers map onto the world longitudinal reference points, which in turn facilitate navigation, exploration, and settlement. Until the 1760s, it was nearly impossible to determine longitude at sea, because pendulum clocks had trouble running regularly on board ships. And even on land, calculating longitude was no small feat; just an estimation of it required a well-regulated clock, astronomical instruments, and observatory charts. Before the use of the telegraph to ascertain longitude in the 1850s, most American communities followed a standard of time that was not necessarily their locale's true time. Precise local time was elusive; for the most part clocks and bells conveyed estimations of the "true time." Time measurements were then, as they are still, a social convention, a technological gauge, rather than an expression of natural law or scientific truth.[23]

As a social convention, time measurements typically are bound up in contests for power and authority. When there was only one clock or bell, it was easy for communities to cohere around its time as their standard, but discord ensued as to whose time counted once there were multiple public clocks. Entangled in this competition were claims made about the accuracy of time-

Sundial, ca. 1800, item X.115, Luce Center, New-York Historical Society. The octagonal copper sundial belonged to Christopher Colles, 42 Pearl Street, New York, identified in city directories as an engineer.

keepers. The clock's sponsor and keeper might enhance or undermine its timekeeping reputation. Ultimately in terms of setting local time standards, the aural and visual reach of the clock mattered as much as, if not more than, its perceived accuracy.

It is important for our understanding of nineteenth-century clock time to distinguish between precision and accuracy.[24] A precise clock computes and records the passage of time (in terms of seconds, minutes, and hours) consistently. A clock set to the wrong time can still be precise if it measures out each second, minute, and hour correctly. An accurate clock, however, is one set to the correct time (as determined according to the prevailing standard).

When the sun was the standard, anyone could easily check a clock against it to determine its accuracy. If a clock kept mean solar time, as we have seen, an equation of time assisted in this exercise. Clocks set to other time standards, such as sidereal time measurements taken from the movement of the stars, required astronomical instruments and specialized knowledge in order to ascertain their accuracy.[25] Once set to the correct time, a timepiece will depend on precision clockworks to maintain its accuracy.

Through the eighteenth and into the nineteenth century, in all but a few places a clock was said to be accurate if it was in accord with the sun. But as public clocks proliferated, their keepers followed different conventions when setting them: some used sun time, others mean solar time, and the real specialists opted for sidereal time. The public did not necessarily seek out clocks showing its preferred time standard (solar, mean solar, or sidereal), instead it deferred to the time indicated by the area's most prominent clock or bell. Prominence could result from material properties like size, beauty, sound, or from social ones like political authority or economic power, or from geographic ones like centrality, visibility, and audibility. The 1852 and 1854 timetables for Lowell Mills stipulated different clocks as the standard for operations. First the time "shown by the regulator clock of JOSEPH RAYNES, 43 Central Street" was the standard, but then AMOS SANBORN convinced the mill managers of the superiority of his regulator. In 1864, Lowell's timetable left aside local jewelers, indicating instead that "STANDARD TIME will be marked at noon by the BELL of the MERRIMACK MANUFACTURING COMPANY."[26] By midcentury, the clock and bell had gained vernacular status as a source of time, rather than just as an expressive vehicle meant to show the time. Time specialists like jewelers, clockmakers, and astronomers continued to look to the sun, the stars, and longitudinal measurements to bolster the authority of "their" time as shown on timepieces under their charge.[27] Everyone else looked to clocks for the time. The mandate of a local boss, confidence in the local jeweler, faith in city government, or the constancy of a church's clocks and bells were among some of the forces that might have shaped preference for one public clock over another. As the size and scale of communities grew during the century's second half, the necessity of accuracy increased because no single clock was authoritative on its own. By the 1870s, public clocks operated within a network of clock time.

Even if public clocks were set accurately, most lacked precision. Their imperfect timekeeping mechanisms and the exposure to which many of them were subject caused clocks to lose or gain time until they no longer displayed accurate time. Particularly when clocks and time balls malfunctioned, the constancy of mechanical time came into question. Until the 1880s, the American temporal system, if it could be called that, was built upon unsynchronizable and imprecise timepieces, in addition to indeterminate time standards. As a consequence, clock time during its formative years was chronically unstable. Its attractions, however, were formidable. As a working system of standard time, the means for accurate time distribution, and a multitude of precise timekeepers emerged, they together enhanced the authority of clocks and of clock time. Clocks known for precision, which mostly belonged to railroads, governing institutions, and jewelers, gained legitimacy because they were more likely than others to display accurate time. But even with standards set and clocks improved, clock time remained an unstable category, one whose constancy all desired, but whose unreliability nevertheless was an important characteristic to which this book attends.

During most of the nineteenth century, then, neither natural nor clock time had complete authority. In this context, pocket watches were reassuring. The high-quality ones, known as "chronometers" or "pocket chronometers," could keep the time far better than most public clocks, with their cumbersome works and pendulums. To be sure, many pocket watches were "delicate machines," as one period repair manual described them, that broke of their own accord as well as in the hands of their owners.[28] When the sun set the standard for the time, watches set to solar time were available for consultation on cloudy days and after dark. Watches made time portable—time specialists used them to carry the time from the site of its measurement and calculation to the public clocks they regulated. Even after the innovations of telegraphic and electric synchronization eased the task of setting and regulating clocks, well-running pocket watches were still used to adjust public clocks.[29] Norman Rockwell's nostalgic painting *The Clock Mender* depicts its eponymous protagonist using his pocket watch to adjust Marshall Field's clock. When in late 1945 Rockwell's rendering of its famous clock appeared as the cover of the *Saturday Evening Post*, the Chicago department store defensively pointed out that its clocks were set by signals sent from the US Naval Observatory.[30] By the 1940s, pocket watches were anachronistic, signs not of precision and accuracy, but of bygone days when clocks ran awry.

While harkening back to older traditions associated with clocks in its design and decoration, the Jewelers Building clock was every bit as modern as its automobile elevator. The master clock at the Elgin factory's astronomical observatory, a precision timepiece known as a "regulator," telegraphed a time signal that kept the elaborate clock accurate to within one-hundredth of a second.[31] The introduction of the possibility of precision to horology several centuries earlier had been an important step toward modern time awareness. Christiaan Huygens's addition of a pendulum to clockworks in the 1660s

Norman Rockwell, *The Clock Mender*, cover for the *Saturday Evening Post*, 1945, oil on canvas. Printed by permission of the Norman Rockwell Family Agency Book Rights Copyright © 1945 the Norman Rockwell Family Entities. Reproduction of painting courtesy of the Chicago History Museum.

enhanced the precision of clocks; by the 1680s, the escapement and balance spring bolstered the precision of watches too. Well-built and well-maintained timekeepers could now track regular intervals of time, whether minutes or hours. Most faltered in their duties, but the ideal of regularity was inscribed as hour and minute marks upon the dials of clocks and watches.

These horological innovations introduced "a temporality of smallness and sameness," according to literary critic Stuart Sherman, in which "regular intervals" became a common feature not only of mechanical timekeepers but also of diaries, periodical literature, and newspapers. In his view, the "blanks" between the marks on the clock dial constitute the primary feature of modernity: that is, they represented a reproducible unit of measure that could be filled in an infinite variety of ways. Sherman shows how time became a matter of measure, rather than of occasion.[32] The form of time kept by early modern clocks and watches precipitated a profound shift in time consciousness among the elite who pondered them. There is much to recommend Sherman's interpretation, for it attends closely to how form—of tools, of spaces, of texts, of images—shapes consciousness and practices. In doing so, it pinpoints one of the ways that clocks wrought a fundamental change in temporal orientation. Nevertheless, conditions in early modern Europe had not caught up with clocks' capacity to modernize social and economic relations: clocks still operated in communities bound by geographic, temporal, and productive constraints. Distances, still traversed with human, animal, or wind power, remained great. Diurnal and seasonal rhythms continued to profoundly differentiate night from day. Economies of scarcity dictated the habits and practices of the vast majority of people. To wit, the world was not yet modern.

Modern it would become when time and space ceased to pose meaningful barriers to the circulation of news, ideas, goods, and people. When it took a long time to traverse even small distances, the differences in time between places did not much matter. Each place had its own time, set by the sun and the stars. But the speed of railroads, steamboats, and telegraphy knit larger and larger geographic expanses together into one place. Steam engines, gaslight, telegraphs, telephones, electricity, phonography, photography, combustion engines, and wireless communications fundamentally changed social and cultural experiences of time and space. Ever increasing speeds—of mechanical operations and of the circulation of goods, capital, and people—necessitated ever closer attention to coordination, synchronization, and precision in timekeeping. In reducing the length of time it took to relay a message or travel to a destination, communication and transportation technologies served to compress, though not eradicate, geographic space. Although the importance of distance receded, that of duration only intensified: the less time it took to

Trade card for watchmakers and jewelers Farr and Olmstead, ca. 1870, Landauer Collection, New-York Historical Society. This advertisement shows Father Time, behind the wheel of a steam engine made of Elgin watches, tossing away useless watches that have taken the shape of livestock—Swiss watches, "Bogus" watches, "Imitation" watches, and even watches that arrive "C.O.D."

communicate or travel, the more emphasis on the time kept by clocks and watches.[33] In this book, I refer to the set of conditions that required constant, precise, and standardized timekeeping, and to the consequential emphasis on speed, urgency, contingency, and simultaneity, as "modernity."[34]

The temporal demands of modernity elicited a range of responses from artists, scientists, and entertainers, most of which are beyond this book's purview. The dissonance between subjective temporal experiences (often called "private time") and objective measures of time (referred to as "public time") animated the early-twentieth-century artistic movement of modernism. The European cultural historian Stephen Kern explains how its adherents developed new narrative and visual forms to explore the disjuncture between public time and private time. Tensions arising out of the clash of national and global time made manifest by the Greenwich Observatory's status as the prime meridian inspired modernists in Britain.[35] The profusion of synchronized public clocks in Bern awakened the young Albert Einstein's recognition of relativity; miniscule measures of time, like a tenth of a second, captured the attention of physicists and philosophers.[36] In the commercial arts, filmmakers experimenting with cinematic narrative began to compress and expand time through editing of sequences.[37] The influence of clocks and clock time on writers, artists, and scientists is well documented and richly substantiated; it makes a persuasive case for accounts of modernity that situate technological change within broad cultural contexts.

The temporal demands of modernity also resulted in changed timekeeping practices. Under modern conditions, the complexities of attending to local time accumulated, leading to successful efforts to dispense with it altogether. In late-eighteenth- and early-nineteenth-century Europe, the lines of temporal authority led to master clocks set by law to the prevailing time standard.[38] But in the United States, no single authority asserted control over the time. Various local institutions competed for this distinction by installing prominent clocks and by drawing attention to the scientific methods or services they used to set their regulators. Over the course of the nineteenth century, conventions whereby localities across a region followed the same time were implemented in Europe, Britain, and the United States. These "standard times" further dissociated clock time from natural time. They severed time from the local and the particular. Without the local and particular, clock time became even more tightly tethered to institutions, procedures, and standards. It depended on reliable clocks and firmly established time standards.

Like Father Time with his reaper atop the eight-ton Jewelers Building clock, time has always disciplined humans. Time begins with sunrises, births, arrivals; it always ends, in a sunset, a death, a departure. The shape of time varies from culture to culture and age to age, as do the ways of measuring and marking it. "Time discipline" is shorthand for how time—ideas about it, ways of measuring it, instruments for meting it out—controls actions, thoughts, dreams, desires. Every culture has its own form of time discipline, its own ways of marshaling ideas about time and methods of measuring it that together generate sanctioned behavior. The social historian E. P. Thompson drew attention to time discipline in a 1967 landmark essay about the English working classes. He set the terms of debate about time discipline for the next quarter century or more, linking it by implication with factories and clocks. Today, owing to the efforts of scores of anthropologists and historians, a much richer theoretical and historical understanding of the concept is in place than the one Thompson deployed.[39] In this book, "modern time discipline" refers to a cultural system in which clocks provide the reference point for temporally organizing and controlling society.

Clocks gained ruling authority, as Michael J. Sauter explains, only after "people stopped disciplining clocks." Before Western Europeans would submit to clock time, the vast majority of a community's timepieces had to show correct time (that is, be accurate) and had to keep time reliably (that is, be precise). It was necessary for there to be confidence that the time shown on a community's master clock, as well as on all the clocks and bells connected to it, was the correct time. To Sauter, a historian of ideas, modern time discipline was the consequence not of industrialization, as Thompson insisted, but of Enlightenment science.[40] In the United States, scientists, clockmakers, government officials, and business managers seeking order and efficiency ushered in modern time discipline.[41] As ordinary Americans accumulated personal timekeepers, financed the installation of public ones through lotteries, gifts, and taxes, and evinced fascination with and obedience to clock time, they too added impetus to the quest for synchronized clocks. Precise clocks indicating accurate time coalesced at various periods in various places across the United States, but overall it would be fair to claim that this feature of modern time discipline was in place by 1900.

Coordinated and well-running public clocks made it possible for modern time discipline to become a fact of American life. During the nineteenth century, many Western Europeans and Americans organized their lives according to clock time, though they did not entirely jettison fealty to natural time indicators like "sun rise." It was the public clock presiding over the street, not E. P. Thompson's factory clock, that captured the imagination and earned the trust of what Sauter has called "a clockwatching public."[42] If far from cities themselves, Americans encountered public clocks in print, where newspapers, periodicals, and books wove them into their accounts.

Unlike rural populations elsewhere, Americans "succumbed to clock time," as the historian Mark M. Smith puts it, because all sorts of clocks were everywhere.[43] Public clocks distributed the time as well as the ideal of "the time" to all who cared to look or listen. Surprisingly, the importance of clocks and bells in American civic spaces, the contests for authority they triggered, and the dreams of synchronicity that they represented have all been overlooked.[44] These timekeepers, even if imperfect in their operations, as they often were, provided an organizational conceit and procedure whose force gradually accumulated with each decade of the nineteenth century.

As clocks saturated public and private spaces, habits and expectations shifted away from a form of time discipline referred to as "time obedience." Communities who lived with few if any clocks, like colonial settlers, antebellum plantation slaves, or prisoners, were time obedient. To greater or lesser degrees, they were responsive to the cacophony of audible and visible signals indicating when to begin, shift, or end an activity. As communities became too large, too dispersed, and too independent to control with audible time signals alone, their members were expected to marshal their own time through the consultation of clocks. To paraphrase Thompson, modern time discipline depends on inward notation or apprehension of the time. Clocks "facilitated a change," as social theorist Barbara Adam explains, "from obedience to the public call of time from the bell tower to internalized time discipline."[45] Clocks and watches helped individuals track the passage of time down to the minute, coordinate with schedules and timetables, and achieve time efficiency. They placed at a premium the anticipation of the arrival of a particular moment of time. Clocks disseminated "an abstract system" that, in the words of anthropologist E. E. Evans-Prichard, produced "autonomous points of reference to which activities ha[d] to conform with precision."[46] Clock time, then, was a handy autonomous point of reference for a people seeking time discipline. It provided the information through which temporal anticipation could take shape and guide actions.

Not only did the ubiquity of clocks contribute to the diminished importance of time obedience, it enhanced the possibility of time orientation. For a number of reasons summed up in the words "industrial production," the value of labor came to be measured in increments of time, rather than by the tasks completed. The division of labor and use of machinery transformed production into a series of steps performed by deskilled laborers and exact machinery; the task orientation of artisanal manufacturing waned. In the factory system, hourly wages and the imperatives of efficiency—maximum output per hour worked—require fealty to clock time. Work became time oriented. Here too clocks play a pivotal role. They provide a consistent means to measure increments of time; ideally the duration of five minutes is exactly the

Elgin National Watch Company plant, Elgin, Illinois, 1914, Haines Photo Company, PAN US GEOG—Illinois no. 35, Prints and Photographs Division, Library of Congress.

same regardless of place, task, or clock. Increments of time are thus akin to other interchangeable parts at the core of mass manufacturing techniques.

Like Elgin's watch factory with its enormous clock tower sporting dials nearly fifteen feet in diameter, factory clocks were large and imposing, dominating space well beyond factory gates.[47] Nevertheless, only 5 percent of public clocks installed between 1870 and 1910 were on the exterior of factory buildings (see table 2). Two types of clocks typically found inside factories like Elgin's, the watch clock and the time clock, warrant some discussion. During the last third of the nineteenth century, many factories, warehouses, and office buildings installed watch clock stations throughout their premises to ensure that watchmen made their rounds consistently; this was especially important in the task of establishing the whereabouts of watchmen in the all-too-frequent cases of fire, theft, and vandalism. While in some cases the regulator for a watch clock system might have been placed in a public space, such as the front office, in most cases these clocks did not disseminate the time.[48] Time clocks (also known as "punch clocks"), in contrast to the ancient origins in "the watch" of watch clocks, have relatively recent origins in the late nineteenth century. In the 1880s and 1890s, time-recording clocks, timecard systems, and automatic time stamps were introduced; by the 1900s, the International Time Recording Company, which would become International Business Machines (IBM), had consolidated the industry. By 1910, nearly every factory and many offices in the United States had a time clock. It is claimed that nearly one hundred million time cards were sold annually during the 1930s.[49] Time clocks narrowed the meaning and domain of clock time into a measure of commodity relations. They were never decorated, never installed in grand public spaces, but instead were furtive, ugly mechanisms hidden in entryways and coatrooms.

Public clocks, unlike those installed in factories, engendered a sense of the common and equitable ownership of time; obscured was the fact that time was a contested resource over which some people had more control than

Table 2. Customers for Seth Thomas and E. Howard clocks, 1871–99, 1902–11

Customer	Number	Percentage of Howard and Thomas business
Government agency	1,006	32
City government	(321)	(10)
Federal government	(259)	(8)
County government	(234)	(8)
State government	(70)	(2)
Unidentified branch	(122)	(4)
Commercial agency	738	24
Jewelers	(178)	(6)
Banks	(106)	(3)
Church	354	11
Other	248	8
Railroad	175	5.5
Factory	151	5
Export	155	5
Private schools and colleges	119	4
Unidentified customers	172	5.5
Orders placed by dealers for unnamed customers	(170)	

Source: Based on the account books of the Seth Thomas Clock Company, held by the National Association of Watch and Clock Collectors and the account books of the E. Howard Clock Company, held by the Archives Center, National Museum of American History, Smithsonian Institution. $N = 3,118$ orders.

Note: Table 2 draws on 1,994 of the 8,149 orders in the extant account books of the E. Howard Company covering the years 1871–1905 and on all 1,124 orders in the extant account books of Seth Thomas covering the years 1877–93 and 1903–11.

others. Public clocks elicited many responses, including the famous antics of Harold Lloyd, who in the moving picture *Safety Last* (1923) dangles from the hands of a public clock far above the street, then momentarily nests in its clockworks like a bird, before his feet get tangled up in the springs. In the feature's final stunt, Lloyd becomes a human pendulum as he swings head first from a rope anchored to the skyscraper's roof. For too long our attention has dwelled on Fritz Lang's *Metropolis* (1927) and Charlie Chaplin's *Modern Times* (1936), where factory clocks are dehumanizing forces keeping capitalists' time. And well they might have been within factory gates, but beyond the shop floor timepieces more often worked in concert with individual Americans than in opposition. Although other temporal orientations rooted in diurnal, seasonal, biological, religious, and historic measures persisted, clocks naturalized the hegemonic use of hours and minutes to measure time.[50] Public clocks and personal timekeepers together operated within social and cultural settings in which Protestant notions about the value of time pulsed, pushing Americans to strive to "redeem the time" through greater work effort, improved efficiency, and a sustained suspicion of idleness and waste.[51] Infusing public clocks with national symbols, especially those associated with the nation's founding, reinforced a commonsense understanding of time as everyone's patrimony. Modern time discipline depended on this sense of clock time as impartial and transcendent, like the weather. Only public clocks could make the case that this was indeed so.

Marking Modern Times draws on the history of timekeepers, especially their manufacture, marketing, and acquisition, to chart certain points along the trajectory toward modern time discipline. Clock time, it shows, was the result of procedures, practices, and assumptions that severed time from nature, the local, and the particular. During the first few decades of the nineteenth century, clock time was an unstable property, subject to scientific mistakes, mechanical vagaries, and unclear time standards. Even through the Civil War, the absence of time standards and reliable techniques for distributing them contributed to the instability of clock time. By the 1890s, clock time had shed its indeterminacy; it was the most-used temporal measure among people of all classes and origins. The precision of timekeepers had improved, an agreed-upon system of time standards was in place, pocket watches were widely distributed, and spectacular timepieces commanded attention. Clock time was irrefutable, so much so that once an effective means for disseminating accurate time over the wires and airwaves (through telegraph, telephone, and radio communications) developed, public clocks themselves could be dispensed with.

My first chapter begins in Philadelphia in 1752, with the arrival of a bell from London; it then describes the gradual accumulation of mechanical timekeepers over the course of the Revolutionary era and early decades of the Republic. Chapters 2 and 3 continue the story through the middle decades of the nineteenth century, introducing jewelers, mechanics, and artisans, as well as "clockwatchers" and watch aficionados. Chapter 4 explores how the process of standardizing time enhanced the authority of public clocks. Chapter 5 sets the public ownership of clocks in the context of the history of governance and timekeeping. Chapter 6 brings the book full circle, returning to Philadelphia in 1898, where the Old State House's clocks and bells stood across town from the city hall's tower clock, which at the time of its installation was the largest in the world. The book closes with an account of the end of the public clock era.

While the jewelers' clock remains on the Chicago building where it was installed in 1928, the jewelers whose offices were within long ago moved away. The modern features of the building, like its indoor parking garage and automobile elevator system, now seem as antiquated as public clocks and pocket watches. The Elgin National Watch Company itself went out of business in the late 1960s; unlike its American competitors Hamilton, Waltham, and Gruen, which all moved operations to Switzerland, a manufacturer in China owns the Elgin brand name. E. Howard and Seth Thomas long ago ceased business; apparently there is no profit to be eked out from their brand names, since no foreign manufacturer has acquired rights to either. Looking back from 1928, this book returns to the epoch when American uses of and dreams about bells, clocks, time balls, standard time, and pocket watches together made and marked modern time.

CHAPTER 1

Time's Tongue and Hands

The First Public Clocks in the United States

From youth to age the sound is sent forth through crowded streets, or floats with sweetest melody above the quiet fields. It gives a tongue to time, which otherwise would pass over our heads as quietly as clouds, and lends a warning to its perpetual flight. . . . It is the voice of rejoicing at festivals, at christenings, and at marriages, and of mourning at the departure of the soul; and when life has ended we leave the remains of those we loved to rest within the bell's deep sound. Its tone, therefore, comes to be fraught with memorial associations.

—William Sidney Gibson "Essay on Church Bells" (1854)

When swung, a bell's clapper, often called its "tongue," strikes the inside of the instrument, which produces a resonant sound; hammers or mallets can also be used to hit the exterior of a bell to produce a sound. Bells themselves give "a tongue to time," that is, they make time audible, transforming it from something that might "otherwise pass over our heads as quietly as clouds" into a distinct sound, the sound of time.[1] The sound of time can also emanate from a horn, a drum, or any instrument, but the sound of bells has exercised particular power in American towns and cities since the colonial period. "A listener hearing both the striking of the hours and the more solemn pauses marking services or ceremonies," as the French cultural historian Alain Corbin observes, "has to cast his response in terms of a double temporal system."[2] It was just such a rich and complex temporality that emerged in Philadelphia during the middle decades of the eighteenth century, and persisted into the nineteenth century.

The casting of bells is a venerable tradition that dates to antiquity in both the East and the West. In Western Europe beginning in the thirteenth century, time's tongue became multivalent; as time became a matter of measure, bells announced the hours, even as they continued to mark occasions.[3] Ringing bells demarcated moments in the arc of individual lives, especially weddings and funerals; they announced sacred times (Sabbaths and other holy days); they rang in honor of important civic events, such as martial triumphs or the anniversaries of the births, deaths, and ascensions of rulers; they called attention to emergencies like fires and floods. In addition to these occasions, bells rang to indicate the passage of time itself (curfew bells, hour's bells, midsummer's bells, and New Year's bells).[4] The historian Richard Rath perceptively comments that bells "mediate[d] between smaller social structures and larger identities (based in religious beliefs, town, region, nation, and colonial relations)." Bells in early America, he explains, "did more than ring out to the heavens; they rang in the state." Hung on the most important buildings and at the highest reaches, bells were "the most expensive ornament of English identity."[5]

The nineteenth century resonated with time's sounds; hour's bells rang throughout American cities and towns, alarms punctuated bedrooms and bunkhouses, gongs and horns dictated action in fields and factories. Catholic church bells rang the Angelus to call devotees to prayer morning, noon, and night. After the 1870s, a few dozen wealthy congregations, mostly in and around New York City, installed big bells to ring the hours and Westminster chimes, which marked each quarter hour with a melody. Clockworks controlled the Angelus bell and Westminster chimes so that they rang at appointed times, thus their sounds contributed to the sense of time as a matter of measure, rather than of occasion.[6] But in the biggest cities by the end of the century, the aural world of public time telling began to fade despite the installation of ever larger bells, drowned out by screeching machinery and crowd noise, or dissipated by wide streets, tall buildings, cavernous alleys. The sound of time never completely disappeared, but the contours of time discipline shifted toward modern practices that depended on a seeing public, on visible timekeepers that were at once accurate and precise, and on what E. P. Thompson described as "the inward notation of time."[7] This chapter explores the early history of public timekeepers in the United States by look-

The Old Liberty Bell, 1899, stereograph, LC-USZ62-77866, Prints and Photographs Division, Library of Congress.

ing closely at the city of Philadelphia, where during the 1820s and 1830s traditional ways of marking public time, particularly reliance on bells, began to give way to modern timekeeping practices and mechanisms.

During the colonial period, bells were considerably more important to public timekeeping than clocks. By 1640, more than two dozen towns in Great Britain had installed ringing clocks on their civic buildings. In the 1650s, Boston and Salem each had ringing bells, as did the New York State House, which as of 1656 rang the hours and an evening curfew at nine. A bell known as "the Province Bell" was hung from the crotch of a tree in Philadelphia in 1685, and in 1697 was moved to the cupola of the newly built courthouse.[8] Settlements in the interior might have been home to a small clock or sundial, but bells, owing to their expense, were out of reach for most of them.[9] In 1713, a clock face was installed on the façade of Boston's new brick state house, and three years later, New York's new state house became home to a "paublic Clock with four Dyal Plates."[10] The fledgling nature of American metal casting technologies, the great expense of importing bells of sufficient size, and the difficulties attendant to hanging bells limited the number and extent of bells. Bell towers of various designs appeared in the early eighteenth century in towns along British North America's Atlantic coast, particularly Charleston, New York, Boston, and Philadelphia.[11] Philadelphia's Christ Church hung a 700-pound bell in 1702 and a much smaller one of 150 pounds in

1711.[12] Gradually the colonies were accumulating bells, which after church buildings and state houses were the most important civic and religious accoutrements of the period.

Shortly after the impressive steeple of the Pennsylvania State House (now known as Independence Hall) was completed, a bell from England's Whitechapel foundry arrived in 1752; almost immediately the one-ton bell cracked. Recast by local brass founders, it was hung the next year with, according to one period account, "Victuals & drink 4 days." Soon complaints were made about the bell's tone.[13] Recast again, the bell was still found insufficiently resonant. This bell, which would become known as the Liberty Bell owing to the inscription "Proclaim Liberty throughout all the land unto the inhabitants thereof," was left in the steeple, and another bell meant to ring the hours was ordered from England. When the second bell arrived, it was installed in the state house tower. It was attached to a clock movement (made by local clockmaker Thomas Stretch) situated in the middle of the building. Rods connected two clock dials under the gables of the east and west exterior walls of the state house to the clock movement. An ornamental case of stone,

John Lewis Krimmel, *Election Day*, 1815, oil on canvas, Winterthur Museum. The Pennsylvania State House's tall-case clock that stood on the building's exterior wall is prominent in this fanciful painting of an election day. Krimmel, a keen-eyed genre painter, included tall-case clocks in some of his other paintings, such as *A Country Wedding* (1820) and *A Quilting Frolic* (1813), in which an ornately cased clock towers in a corner of a bare-floored room.

in the fashion of tall-case clock cases, protected the dial on the west wall's exterior.[14] The state house clock movement, clock dials, and bells worked together to enrich and complicate Philadelphia's temporal rhythms. The new bell rang the hours; the gabled clocks showed the hours and minutes; and the old bell, the Liberty Bell, sounded fire alarms and marked mournful and joyous occasions.

Shortly after the installation of the state house bell, a lottery raised enough money to finish Christ Church's new steeple and purchase a ring of eight bells that together weighed eight thousand pounds, but it did not bank enough for a clock, as had been intended.[15] These bells, like other church bells throughout the colonies, rang at the opening and closing of markets as well as to bring townspeople to church services, in keeping with practices dating to medieval Europe.[16] The ringing of the hours on bells was an infrequent occurrence, owing partially to the rarity of reliable clocks. In terms of timekeeping, as in other spheres, church and state worked together during the colonial period, relying more heavily on bells than clocks to disseminate the time, which typically was the measure of occasions like the opening of a market or the commencement of a religious service.

Neither church nor state had much of a hand in determining the time; that task was left to amateur scientists, jewelers, and clockmakers. Shortly after the installation of the Pennsylvania State House bells and clock, members of the fledgling American Philosophical Society built an astronomical observatory on the house grounds. Although ships had been traversing the world's oceans for centuries, it was only in the 1750s that a reliable method of navigation emerged in the shape of a marine chronometer. Because this precision timepiece did not lose or gain time despite the vagaries of being at sea, it facilitated the calculation of longitude and thus eliminated some of the mysteries of navigation. Now ports and coasts could be charted, their longitudes determined.[17] The half dozen men who constituted the American Philosophical Society wished to make their own contribution to science and navigation. Their observations in 1769 of the transits of Venus and of Mercury from the state house observatory, the colonies' first astronomical observatory, resulted in the determination of Philadelphia's longitude.[18]

Afterward the state house yard observatory was dismantled, and the Philosophical Society's transit instrument was moved to the south windowsill of the state house tower. The observations made with the transit instrument helped determine Philadelphia's official time, mean solar time, which was used to regulate the state house clock.[19] Thomas Stretch, who made the clock, had maintained and regulated it until 1762, after which time the clockmaker Edward Duffield took over those duties. The talented clockmaker David Rittenhouse was already in charge of the Philosophical Society's tall-case regulator clock, which was kept inside the state house. In 1775, he took over the regulation of the state house clock, relying on, as he wrote in his application for the duty, "astronomical instruments for adjusting it."[20] The clockworks attended to by Stretch, Duffield, and Rittenhouse sent Philadelphia's official time to the dials on the state house's tall-case clocks and to its bell and then beyond.

Not everyone in colonial Philadelphia was pleased about having the hours rung on the state house bell. In September 1772, Philadelphians living near the state house petitioned the Pennsylvania General Assembly for relief from "the too frequent Ringing of the great Bell in the Steeple of the State-House." They complained that the bell ringing "much incommoded and distressed" them, particularly when family members were sick, "at which Times, from [the bell's] uncommon Size and unusual Sound, [the ringing] is extremely dangerous and may prove fatal." What is more, the petitioners asserted, the bell was not supposed to ring except on public occasions, such as when the assembly or courts met. The assembly tabled the petition. Shortly thereafter the steeple had grown too rickety to bear the weight of the bells, and no mention is made in historical sources of their ringing between 1773 and 1781.[21]

Despite the interest of the account of distress the bell caused, a complaint not as unusual as might be thought, we should focus on the disavowal of the utility of ringing the hours.[22] The petitioners claimed that announcing the hours served no public purpose except during the brief periods of legislative and judicial sessions. This might lead one to suspect that residents of Philadelphia did not yet have a use for the hours, except with regard to coordinating the meetings of governing officials. Such a conclusion would be misguided. Rather than living in a state of natural time where the hours did not matter, Philadelphians, like other colonials living in British North America, consulted a variety of timekeepers and followed several different temporal systems including clock time. If they wished to know the hour, they looked to their clocks, watches, sundials, or the sun itself. But the arrival of ships, wagons, storms, full moons, nightfall, harvests, and Christian holy days held more sway over their rhythms and routines than did that of noon or midnight.

Archaeological and probate data from British North America show an increase in the ownership of clocks and watches over the course of the eighteenth century.[23] No data come directly from Philadelphia, but other sources support the conclusion that timekeeping was of great interest to the city's residents. In nearby Annapolis, for instance, the incidence of timepieces rose from 6 percent in 1688 to 48 percent in 1777.[24] Notwithstanding a rise in demand over the first part of the eighteenth century, by its end demand for clocks had become somewhat anemic. Not only were they expensive, owing to their metal clockworks, heavy weights, expensive cabinetry, and painted dials, but their pendulums required tall cases. What is more, they required the regular attention of a clockmaker, for unlike today's clocks, they all ran fast or slow, that is, when they did not run down altogether. As a young man, the inventor Eli Terry began making clocks in the 1790s only to discover, as his son Henry recalled, that "the demand for clocks was limited." Henry reports his father making three or four brass movements at a time and then strapping them to his horse so he could "put them up for purchasers." Perhaps he might have built twenty-five such clocks a year.[25]

Through the end of the eighteenth century, ownership of a pocket watch was more likely than that of a clock.[26] Jewelers and watchmakers in colonial cities imported watch movements and parts for assembly, often engraving their own signatures and place names onto the European movements. They sold the cased watches for fifteen to thirty dollars and often much more.[27] In 1750, Benjamin Bagnall advertised that he had received "a neat parcel of gold, pinchbeck [a gold alloy] and silver watches" as well as West India rum, which he was offering for sale in Philadelphia.[28] Joshua Lockwood, a self-proclaimed "watch maker" in Charleston, announced in 1760 that he had imported a variety of watches and clocks, which he would sell at "London retail prices." The decoration of his clocks and watches showed various scenes, including slaves planting a field and a "Cherokee Fight."[29] Watches were less expensive and sometimes better timekeepers than clocks during the eighteenth century. And as status items, they were easier to flaunt than a tall-case clock. "Silver watches signaled enviable affluence combined with a suitably masculine command of technology," which historian John Styles explains was "of the kind richer men manifested through purchases of scientific instruments and the like."[30] Benjamin Franklin's triumphant return from Philadelphia to his brother's Boston printing shop in 1724 included his flourishing a watch: after he, in his own words, "produc'd a handful of Silver" to show the shop's journeymen just how far he had come from his apprentice days, he "took an Opportunity of letting them see [his] Watch."[31]

Silver coin and watch, each tools and signs of success in the colonial metropole, connoted gentility after the Revolution too: a 1785 exchange in the *Pennsylvania Packet* depicts an adversary as "walking the streets of Philadelphia with a white hat, and a watch in his pocket."[32] There was a considerable flow of English timepieces to the newly formed United States, perhaps as many as thirty thousand.[33] The French also took the opportunity to export watches, as well as clocks, to the newly formed nation. One notice from a 1789 newspaper announces "D. F. Lanny, WATCH and CLOCK-MAKER, late from Paris, has brought with him, and received by way of New-York and Philadelphia, a most elegant assortment of WATCHES, CLOCKS and JEWELRY."[34]

In eighteenth-century Philadelphia, as in other cities of that period, pocket watches were extraordinarily special devices above and beyond their expense and technical features. In tandem with the dials of public clocks like the gabled state house clock, they facilitated, encouraged, and privileged the visual apprehension of mechanical time. According to Michael Sauter, "the pocket watch made time a visual phenomenon." In eighteenth-century Berlin, the location of his research, pocket watches "empowered Berliners to monitor and critique their clocks." Sauter gives the name "clockwatchers" to the public who carried and relied on pocket watches to gauge the time. They could not very well consult the clocks in their drawing rooms or studies as they went about their daily business; they needed a timekeeper that was portable. When each week the city's official clock was reset, the clockwatchers adjusted their watches accordingly, perhaps taking the correct time to clocks in their homes and offices. The remainder of the week, they monitored the city clock, relying on the fact that their pocket watches ran with more precision than its large, cumbersome works. Since the public clock kept mean solar time, clockwatchers could not refer to the sun or sundials as their standard. In short, the pocket watch became a standard keeper. Additionally, it "embedded time in daily life by making time-gathering a ritual."[35] No longer was mechanical time a novelty, a fancy notation reserved for births or deaths; it was becoming a feature of ordinary life. As a visual phenomenon, a portable gauge, and a quotidian tool, the pocket watch in concert with public clocks acculturated people to clock time.

Despite the slow encroachments of visual markers of mechanical time, overlapping domains of time remained audible in cities and countryside. Immediately following the first public reading of the Declaration of Independence, from a stage set on the erstwhile site of the Pennsylvania State House observatory, all of Philadelphia's bells tolled. It is unlikely, however, that the two state house bells added to the din, because they had been taken down from the unsound steeple.[36] Nevertheless, a lasting myth was spun that the Old State House bell rang the news, as it most certainly should have given its inscribed command to "proclaim liberty."[37] In mythically ringing, the Liberty Bell announced the nation's birth, a moment John Adams described with the oft-quoted phrase "Thirteen clocks were made to strike together." Some scrutiny of Adams's metaphor underscores certain key points about timekeeping during the Revolutionary and early national eras. First, Adams asks the reader to imagine thirteen clocks audibly demonstrating their coordination by striking together. His was a world in which authoritative time was aurally announced, not read on a clock face. Second, Adam's metaphor was part of a longer sentence, which read in full: "Thirteen clocks were made to strike together a perfection of mechanism which no artist had before effected." It was unimaginable in his day that a mechanism could be perfected such that multiple clocks could operate so well, so precisely, as to keep the same time. The conditions of modern time discipline had yet to arrive, even in 1818 when Adams penned this famous phrase in a letter full of reminiscences.[38] Indeed, they seemed as miraculous as a revolution meant to overthrow a king.

Because of the threat of the arrival of British troops in the early fall of 1777, the Liberty Bell and the clock bell (along with Philadelphia's other bells) were hidden in nearby Allentown. If the British had captured Philadelphia, it seemed likely that they would have confiscated the bells and then melted them down for shot. As in medieval England and nineteenth-century France, bells in colonial British North America bolstered claims to legitimacy and authority, so their capture alone would have symbolically restored monarchical claims.[39] A year later, with the possibility of British invasion receding, the Liberty Bell and hour's bell were brought back to Philadelphia. The Liberty Bell was kept in storage for seven more years because the steeple had gone from rickety to rotten and there was nowhere else for it to hang. The clock bell was hung in a shed jerry-rigged on the roof in front of the tower, where it rang the hours.[40]

The importance of the Pennsylvania State House, its exterior clocks, and the Liberty Bell receded for several decades after the American Revolution, receiving neither public acclaim nor funds. The decayed wooden steeple was taken down in 1781. The one-ton Liberty Bell was reinstalled a few years later, but this time behind the louvers of the tower's middle level. It called voters, opened legislative sessions, celebrated patriotic events, and tolled for the dead.[41] After the Pennsylvania General Assembly moved to Lancaster in 1799, entrepreneurs and civic groups, including the museum operator and painter Charles Wilson Peale, used the Old State House rooms for various purposes, but its clockworks were neglected. Various city papers criticized the clock for its irregularity and Joseph Leacock, its keeper, for his carelessness. Leacock, who was paid by the legislature, defended his reputation in 1800, explaining that he "wound the weight of that Clock full up, expecting it would have gone tolerably regular, but I found it had gained half an hour." The fault he felt was due to the clock's "poorly made" escapement.[42] In 1804, a cupola with clock faces was added to the bell tower atop Philadelphia's Second Street Market (at Second and Pine). Two other public clocks of unknown characteristics and provenance ran, one on High Street, the other on Delaware Street. It was reported that these clocks were "badly put up, and the access to them both difficult and dangerous."[43] In identifying it as "common knowledge" that the Old State House clock ran ten to twenty minutes fast, one 1809 newspaper report reveals that some Philadelphians cared enough about the time it kept to calculate and remark upon its deviations. Its author finger-wagged: "It is very important that this Clock should be kept right—Courts, Schools, Cooks, Coachmen, etc. etc. all regulate themselves by this clock."[44]

The city of Philadelphia did not lack sources of accurate time despite the poor performance of the Old State House clock, but as in the rest of the young nation these sources were not accessible to the entire public. In the last part of the eighteenth century, David Rittenhouse built astonishingly precise clocks for Philadelphia's elite. He also maintained an observatory between 1781 and his death in 1796, at which time the octagonal building became his mausoleum, where a slab of marble with the exact time of his death stood.[45] A handful of clockmakers less talented than Rittenhouse served the local elite throughout the nation's cities and towns. Importers stocked still more clocks as well as watches, as we have seen. Nevertheless, owing to modest levels of wealth and the great expense of timepieces, in 1800 less than a fourth of the American population owned mechanical timepieces of any sort.[46] Although the proportion of Philadelphia's households with clocks may have been higher than elsewhere in the United States, even there mechanical timepieces were not yet everyday paraphernalia. Despite Philadelphia's centrality to American clockmaking prior to 1800, it was in Massachusetts and Connecticut that innovations were being made that would put a clock in nearly every American home, including most of Philadelphia's.

In the 1790s, artisans in Massachusetts, the Willard brothers, introduced the basis for inexpensive small clocks. "Banjo clocks" had simple wooden works, short pendulums, and cabinets that could sit on shelves or hang on walls. The Willard brothers' innovations transformed the manufacture of and market for domestic clocks in the United States.[47] A brief survey of the innovations in the design and manufacture of wall and shelf clocks developing out of the Willard brothers' efforts helps complete the picture in which visual indicators of the time came to complement aural ones, and in which clock time's associations with industry, science, and nation enhanced its relevance for ordinary people. The mass manufacture of clocks stimulated interest in similar efforts to construct watches, though it would only be in the middle of the century that these would come to fruition. But the clocks themselves, in company with public clocks and imported pocket watches, etched grooves within which modern time discipline began to form.

After the Willard brothers' breakthrough, entrepreneurial interest in clocks grew, such that in 1806 Eli Terry signed a contract to make four thousand wooden clock movements in three years for the Connecticut merchants Edward and Levi Porter. A lack of competency with timekeeping mechanisms among the target market is reflected in the fact that these tall-case clocks came with detailed instructions, most of which addressed the clock movements' potential failings.[48] Shortly before the War of 1812, Terry perfected his "Pillar and Scroll" shelf clock, which he patented in 1816. Considered by many "a revolutionary technological and aesthetic achievement," the shelf clock with its ten-inch pendulum, was easy to make, easy to transport, and easy to sell.[49] By 1820 Terry, with the help of thirty workmen, manufactured twenty-five hundred of these clocks a year, in four different styles. Entry costs into the industry were low, so throughout the 1820s other artisans set up manufactories clustered around Bristol and Plymouth, Connecticut.[50] Many of these clocks were painted with scenes depicting Mt. Vernon, George Washington, or naval battles from the War of 1812. The decoration of clocks, both modest American-made ones and expensive European imports, favored national symbols, especially George Washington.[51]

By the 1820s, for the first time, more Americans owned clocks than watches. The social historian Martin Bruegel's aphorism about timekeeping—"people owned timepieces before time owned people"—is both apt and true. His evidence demonstrates that antebellum Hudson Valley households acquired clocks first out of impulses to decorate and furnish their homes, only later in order to keep the time. Before 1820, the prohibitive expense of both clocks and watches reinforced their role as markers of status; by the same token, their relatively sparse distribution undermined their utility. They were ornaments, decorations, embellishments: they connected their owners to what the cultural historian Michael O'Malley describes as "an idealized vision of the world they lived in."[52] Since they were not yet part of a network of time, these clocks did not at first synchronize households with each other, let alone with distant markets. Slowly clocks, connected as they were to notions of gentility and refinement made emblematic by parlors and fine furniture, transcended their status as ornament. They became talismans of orderly households, particularly once householders had access to reliable sources of accurate time, perhaps rung on distant hour's bells or transported on pocket watch dials from town centers. Eventually, once sufficient numbers of Americans acquired clocks, they began to use them to measure productivity, time labor, and coordinate social actions. Pocket watches in this context were tools

Jean-Baptiste Dubuc, George Washington mantle clock, ca. 1805, Winterthur Museum. Produced in France, this mantel clock meant for the American market showcases historic and mechanical time.

for gathering the time. They acted as conduits between public clocks and timepieces shelved on mantels, hung on walls, and sitting beneath stairwells. A network of clock time was emerging.

Around when Eli Terry first began mass-producing domestic clocks, Massachusetts watch assembler Luther Goddard opened the first watch factory in the United States. There he assembled watches out of English watch parts he acquired from a bankrupt Boston importer. Between 1809 and 1817, Goddard sold five hundred finished watches, demand for which was no doubt high owing to the disruption in the trade of English watches following embargoes, tariffs, and the War of 1812. Goddard's enterprise, however, was not especially successful over the long run, so he abandoned it for steadier work finishing and repairing watches in Worcester, Massachusetts.[53] After 1820, watches appear less and less frequently in American probate inventories: whether because of their relative expense compared to clocks or to heirs pocketing them so as to avoid taxes will forever be unknown.[54] Across the United States as a whole, there were far fewer watches than clocks in circulation. This reversal from the eighteenth century, during which time watches outnumbered clocks, is the consequence of the mass production of clocks, for Americans were importing watches and watch works in greater quantities than ever from France, England, and Switzerland.[55]

Philadelphia's business in imported watches, like that of Boston and New York, was robust. Beneath an 1818 engraving of Father Time passing his scythe over the earth topped with a sandglass, the watchmaker Frederick Reed, on Market Street, announced that he had "constantly on hand a general assortment of Gold and Silver Patent Lever and Plain Watches." Jehu Ward's advertisement in the same handbill boasted similar stock. Around the same time as Reed and Ward peddled their watches, navigational instrument maker William H. C. Riggs set up a makeshift observatory to help him in his task of rating chronometers (well-regulated clocks meant to assist in navigation) for captains of ocean-going vessels. He supplied the city, or at least the inhabitants who went to his offices, with "correct time." In addition to other fine instruments, he also sold watches and chronometers. Riggs's business was so good that his firm, Riggs and Brother, became Philadelphia's leading emporium of clocks, watches, and chronometers.[56] Philadelphia with its many artisans and jewelers was also the first center of watchcase making in the United States; a high tariff on imported watchcases helped this domestic industry thrive.[57] Though fewer in number than clocks, imported watches in American cases enhanced the attraction of and potential for a network of clock time emanating from public clocks.

When the city of Philadelphia acquired title to the Old State House in 1818, it embraced the responsibility of maintaining its bells and clock.[58] Increased activity of the courts, growing emphasis on timing in business matters, and rising acquisition of watches and clocks contributed to the city's new attention to mechanical timekeeping. The city's Select and Common Councils appointed a committee to examine the state house's timekeepers in 1821 and

FREDERICK REED,
WATCH MAKER,
No. 150, Market Street,

Has constantly on hand a general assortment of Gold and Silver Patent Lever and Plain Watches, which he will dispose of on the most reasonable terms, by whole-sale or retail.

JEHU WARD,
CLOCK AND WATCH MAKER,

Has taken the stand lately occupied by Abraham Patton,

No. 44, Market street,

(SOUTH SIDE, BETWEEN FRONT AND SECOND STREETS,)

PHILADELPHIA,

Where he keeps constantly on hand a general assortment of Double and Single cased warranted Watches, Chains, Seals, Keys, Silver Table and Tea Spoons, Sugar Tongs, &c. which he offers, wholesale or retail, at reduced prices.

J. W. will also supply orders for all kinds of Silver Ware at the shortest notice and on the most reasonable terms.

1818

Frederick Reed notice in *Paxton's Philadelphia Advertiser* (1818), Warshaw Collection, National Museum of American History, Smithsonian Institution. Reed was not the only watch importer in Philadelphia. Beneath his notice, an advertisement for "clock and watch maker" Jehu Ward announced that he had "taken the stand lately occupied by Abraham Patton."

again in 1828.[59] By the second investigation, it was agreed that, "the time of the citizens of Philadelphia was of so much importance to them, that there ought to be some accurate means of marking its passage." The state house clock, it was lamented, was notoriously unreliable. Indeed, the Clock Committee remarked, "if there is anything proverbial, it is the badness of the clock at the state house." It was called "an *excusing*, not a regulating clock." These complaints were amplified: "It is a clock which affords no rule to go by, but a rule not to go by, for everybody knows it can never go right." Persuaded by these pleas, in 1828 the city's Select and Common Councils appropriated funds to install new clocks and bells on the Old State House and two market houses, and to pay for their upkeep.[60]

As the importance of the Old State House, which was in disrepair, became evident anew to Philadelphians, plans were made not just to spiff it up, but also to restore it to its 1776 condition. The 1824–25 visit of the marquis de Lafayette, a Revolutionary War hero, which coincided with the fiftieth anniversary of the Revolution, awakened a sentimental attachment to the relics from the nation's earliest days.[61] As plans to refurbish the Old State House took shape, the Common Council's Clock Committee proposed purchasing a new clock and bell for $12,000 (about $300,000 in today's dollars). It recommended the improvements to meet "the necessity of having a uniform time for the city," to provide reliable fire signals, and to improve "the appearance of our city, which is so deficient in embellishments, which in other cities are considered indispensable." The architect William Strickland proposed adding a brick turret that was sixty feet higher that the original wooden one, largely so it could bear the weight of a new 4,000-pound bell.[62] Although some objections were made that forced the design to more closely resemble the original steeple, in general "the propriety of erecting a clock upon the State House steeple," a departure from the original design, was agreed upon. The talented artisan Isaiah Lukens built the clockworks for the tower clock. It was hoped that the turret would be high enough that the bell could "be heard over the city," that the new clock dials affixed to the tower would be "seen in distant parts of town," and that the gaslight illumination of the dials would make the time visible even after dark.[63]

The new clock and bell were set in motion New Year's Day 1829. Responses varied. It was Philadelphia's first tower clock, but it lacked the visibility some residents expected. One Philadelphian complained that "two squares distant [from the clock tower] it is mortally impossible to observe anything more than hieroglyphics."[64] The old clock and bell that had rung the hours were consigned to St. Augustine's Catholic Church, in the northeastern part of the city, where it was reported that there was "not a single standard for reference" to the time, nor "the sound of the church-going bell."[65] A few years later, New York added a four-faced clock beneath the cupola on its city hall, and Simon Willard installed one of his tower clocks on Boston's former state house (Faneuil Hall) and another one on Boston's Commercial Wharf Building. After a halt in business owing to the economic collapse in 1837, various small concerns, such as Smith's Clock Establishment in New York City, picked up the pace of their business in the 1840s. Many of the clocks installed in the 1830s and 1840s had striking mechanisms, so their works monitored the striking of hours on bells and gongs as well as the operation of minute and hour's hands on what were mostly in those days painted wooden clock faces. A predilection for tower clocks was evident; as American towns and cities gained modest amounts of wealth, they invested in signs of order, prestige, and permanence. They did so speculating that in being "seen from more than one thousand doors and windows," their tower clocks would establish a time standard for the community.[66] Loud hour's bells also served such aspirations. Time as a matter of occasion was still important, but as a matter of measure it was gaining both visibility and audibility.

When Philadelphia's new tower was completed, the Liberty Bell was

Hard Times Token, 1837, New-York Historical Society. Between 1833 and 1843, small copper coins called Hard Times Tokens circulated as informal currency with advertising messages engraved upon them. The obverse of this one reads "Smith's Clock Establishment, No. 7½ Bowery New York, 1837." Smith sold tower clocks and other timekeepers.

rehung. It only rang on national occasions, such as July Fourth (1831), George Washington's birthday (1832), and to mark the deaths of Lafayette (1834), John Marshall (1835), and President William Henry Harrison (1841).[67] Since for most if not all of its tenure, the Liberty Bell did not ring the hours, its significance as a temporal symbol is within the realm of historic time. When it was tolled in the honor of deceased public figures (presidents, jurists, and founding fathers), it announced the end of epochs. When the city ordered its ringing to celebrate Washington's birthday or July Fourth, it heralded the beginning of an epoch, a renewal of the commitments and bonds that had sent the colonies on a path toward nationhood. Erecting a massive clock tower, installing four clock dials meant to be seen across the growing city of Philadelphia, and hanging a large bell meant to ring the hours on a building coming to be known as Independence Hall suggest that by the end of the 1820s Americans envisioned the conquest of time as much a part of national destiny as the proclamation of liberty throughout the land.[68] But without agreed-upon time standards, let alone methods of coordinating the great variety and number of timekeepers, clock time's authority would remain tenuous.

CHAPTER 2

Clockwatching

The Uneasy Authority of Clocks
and Watches in Antebellum America

He halted for a few moments in order to ascertain the difference in time, between his gold-repeater and the State House clock, which had just struck one. While thus engaged, intently perusing the face of his watch by the light of the moon, a stout middle-aged gentleman, wrapped up in a thick overcoat, with a carpet bag in his hand, came striding rapidly across the street, and for a moment stood silent and unperceived at his shoulder. "Well Luke—is the repeater right and the State House wrong?" said a hearty cheerful voice.

— George Lippard, *Quaker City* (1845)

The contests for authority among antebellum timekeepers cannot be underestimated. Consider the account of the tensions between the state house clock and the pocket-watch carrying public found in George Lippard's sensational 1845 novel, *The Quaker City*.[1] In the passage that stands as the epigraph to this chapter, all the elements of midcentury time telling are evident. The aural and visual compete—the clock has "just struck one," but Luke "peruses" rather than listens. The mechanical and the natural comingle—it is the "light of the moon" that illuminates the "face of his watch." Which is the authority for the time, the pocket watch or the state house clock, the bell striking one or the moon? The passage provides no clue as to which is more reliable, but it is a foregone conclusion that there will be "[a] difference in time, between his gold-repeater and the State House clock." Indeed, every timekeeper seemed to show its own time; there was little possibility of synchronicity, not between aural and visual time indicators, not between natural and mechanical ones, not even between a pocket watch and the state house clock.

One scholar has described the scene from *Quaker City* as "a decidedly urban moment that speaks to the anxieties of a nation that had come to regard its cities as so many undecipherable texts, full of shifting and misleading signs."[2] Clocks were among the most visible of these enigmatic texts. They were found in public spaces among signs, trade cards, handbills, posters, parade banners, newspapers, and currency, all of which constituted what historian David Henkin calls the core of "urban modernity in the capitalist West."[3] But with only a few timekeepers deserving the commendation of "precision," and with accurate time mostly out of reach, responses to these timekeepers in the 1830s and 1840s often included suspicion and mockery. Period fiction expressed reservations about the effects of investment in clockworks and timekeeping. For instance, in *Empire City* (1850), George Lippard described a New York City slumlord "whose secret depravity went by clockwork," the temperance author T. S. Arthur mocks time discipline in "The Punctual Man" (1854), and popular literature and conduct manuals included warnings about "confidence men" using bait-and-switch tactics to steal watches from unsuspecting victims.[4]

The most arresting expression of derision about the expanding domain of clock time can be found in an odd story titled "The Devil in the Belfry" penned by Edgar Allan Poe shortly after moving to Philadelphia in 1838. It describes carved woodwork of "cabbages and time-pieces" that could be found throughout "the finest place in the world," a town known as "Vondervotteimittis" (Wonder-What-Time-It-Is). On the steeple above the town council's chambers was a "great clock" with seven "large and white faces" with "hands heavy and black," which were "readily seen from all quarters." The clock "was never yet known to have anything the matter with it"; indeed "the bare supposition of such a thing was considered heretical." All the town's clocks, watches, and bells were precise and accurate. With the satire Poe was known for, he made it clear that a synchronized town may have been the aspiration of many but was actually preposterous enough for a Gothic tale. In keeping with the tone, at five minutes before twelve a small, dandified man danced through town (without "*keeping time* in his steps"). With half a second left before noon, Poe writes, "the bell was about to strike, and it was a matter of absolute and pre-eminent necessity that every body should look well at his watch." How closely Poe had observed the habits of city dwellers anxious to regulate their watches. The fiddler, seated on the belfry man, rang the hours with bell rope in his teeth. When the clock struck thirteen, the townsmen "turned pale," the other clocks "took to dancing" and striking thirteen, and the town's pigs and cats (who all had watches tied to their tails) went wild. The story's narrator ends the account with "an appeal for aid to all lovers of correct time."[5]

In this chapter, I explore how "lovers of correct time" supported the installation of ever more bells and clocks in American cities, the founding of astronomical observatories, the efforts of the US government to supply accurate time through its own observatory and a novel device known as a "time ball," and the manufacture and distribution of inexpensive household clocks and pocket watches. In doing so, I sketch a picture of the accumulation of timepieces both public and private during the middle decades of the nineteenth century. Running throughout the account is an emphasis on the absence of synchronicity, on the distrust it generated, and on concomitant efforts to secure it. To achieve lasting relevance, public clocks required a single standard of time, a preponderance of pocket watches with which to extend the time, and a "structure of feeling" predisposed to clock time.[6] As I show in this chapter, by the end of the US Civil War (1861–65) the technology, business, and culture of timekeeping were firmly headed in those directions.

Clocks set time standards in Philadelphia and other antebellum cities. Attentive to the fact that work and commerce were becoming time oriented, and that more and more people owned timepieces, a local ordinance required the city of Philadelphia to provide accurate time for the regulation of public timekeepers. To meet this demand, in 1835 the Committee on Public Clocks recommended that public clocks, "be designated by ordinance and so located as to accommodate our Citizens in different Sections of the City." It proposed installing two new clocks to augment the clocks on the Old State House, St. Augustine's Church, and a market house at Second and Pine Streets. In keeping with these recommendations, the city assumed responsibility for the upkeep of the clock and bells in St. Augustine's Church and installed a clock on the Jersey Market.[7] Through the 1830s, the city of Philadelphia maintained four public clocks. Among them, the one on the Old State House remained dominant. Indeed an 1836 guide to etiquette instructed, without apparent irony, that "in Philadelphia it is necessary to be punctual to a second, for there everybody breathes by the State-house clock." You must, it warned, be at your appointments "at the instant the first stroke of the great clock sounds." It further admonished that it would

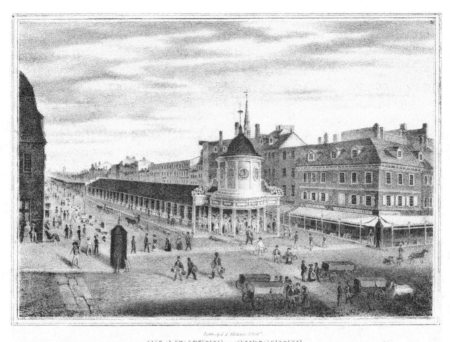

MARKET STREET,
from Front Street
Publ'd by J. T Bowen at his Lithographic & Print Colouring Establishment...N°94. Walnut Street. Philadelphia

J. C. Wild, *Market Street from Front Street*, ca. 1840, lithograph, courtesy of the Library
Company of Philadelphia. During the 1830s, the city of Philadelphia added a clock
adorned with cornucopia to the cupola punctuating one end of the open arcade of the
Jersey Market.

be "useless to plead the evidence of your watch" and that "the unpunctual is
pardoned by no one."[8]

Despite the seeming preference for state house time, the city's clocks
tended to indicate several different times. Grappling with the difficult problem of how to synchronize them, the city appointed an "Observatory Committee" to consider how to enact the ordinance "providing for the regulation
of time keepers." In 1835, the committee suggested moving its observatory
(which at this point was simply a collection of instruments) to a central location where the local time of Philadelphia could be determined and displayed
on a visible public clock. The city fathers did not support this plan.[9] Instead
of observatory time, the city's clocks showed the time as determined by their
keepers. Each clock had an attendant: Isaiah Lukens was paid $100 a year to
regulate the state house clock, which he had built in 1829. Lukens owned

a "Fine Regulating Clock, for Astronomical Observations," as well as other instruments for determining and carrying the time, which ensured that he could reliably fulfill his charge.[10] The city also paid the maker of clocks and watches Benjamin Clark $120 to build and then look after the Jersey Market clock, and the jeweler and watchmaker Charles G. Borhek $40 a year to do the same for the Second Street Market clock.[11] Clark and Borhek likely had use of transit instruments with which to determine the time. Sources do not say how often the clockmakers reset the city clocks, but the common practice was to do so once a week.

As in other American cities of the period, Philadelphia's inventory of public clocks grew over the course of the 1840s and 1850s. Across the nation, churches, civic authorities, jewelers, and newspapers installed public clocks; agents like Thomas Bailey sold tower clocks to southern communities; public clockmakers secured firm footholds in midwestern cities like St. Louis and Milwaukee; and at least six firms competed for local and national contracts out of offices in New York City.[12] In 1842, Philadelphia's most successful newspaper, the *Public Ledger*, mounted a large clock on the corner of its new building at Third Street and Chestnut. A Greek revival commissioner's hall with a prominent clock tower was finished in the Spring Garden district in 1848.[13] Jewelers' clocks were prominently displayed in windows and shop interiors.

Measure, as represented on clock faces and signified by alarms and bells, was gradually becoming the dominant mark of time, but occasion maintained a place, particularly through the tolling of church bells. By the 1850s, three prolific American bell foundries' businesses were well established. The Meneely Bell Company in upstate New York began operations in the 1820s; it may have installed as many as 65,000 bells over the course of its operations. In the 1840s, Ohio's Verdin Company started up; still in business today, it claims 50,000 bells as its output. Outpacing both of these bell foundries, however, is Maryland's McShane Bell Foundry, which estimates it has cast 300,000 bells since 1856. With each decade, these accumulating numbers of American bells have also grown larger and heavier.[14] More than any other purpose, their utility has been in sounding the hours.

The Mechanics' Bell that shipbuilders hung in the early 1830s in New York City's shipyards exemplifies the transition from natural to clock time and from time as a matter of occasion to time as a matter of measure. As George McNeil, an early labor leader and historian, explains: "When the bell was first erected, there were no regular hours of labor, and the mechanic worked from sun to sun" during the long summer days. After shipbuilders won the right to limit their workday to ten hours, they "determined to have a new bell," which they called the "Ten-hour bell." They had an old work bell recast, and as the story goes, they enlarged it with their own "gold, silver and copper coins" contributed to the foundry. On the skeleton tower where it was hung, they erected a sign with the words "Mechanics' Bell." The shipbuilders appointed a "veteran of 1812" to ring the bell four times a day. According to a later report in *Harper's*, they "were insatiable in their demand that [the bell-ringer] should be prompt."[15] Mills and factories that hired wage labor, as well as plantations and farms that relied on enslaved labor, employed bells, clocks, and watches to command punctuality, to measure efficiency and productivity, and in the case of free labor, to mete out hourly wages. Contests over the authority for time within work places frequently flared up, about which much has been written.[16] Like many other period bells, the Mechanics Bell was compared to the Liberty Bell: "As the 'Liberty Bell' rang out the proclamation of liberty from monarchial control, so the 'Mechanics' Bell' proclaimed liberty of leisure for the sons of toil."[17]

The Liberty Bell itself underwent a transformation during the 1840s and 1850s. As it was being rung in honor of George Washington's birthday in 1846, a fracture deepened into a crack, rendering it impossible for the bell to ring. "Left a mere wreck," as one commentator lamented, it remained hanging until 1852, when it was moved to the same room in which the Declaration of Independence was displayed. Around the same time, abolitionists appropriated the bell as a symbol, titling one of their monthly publications

Charles Vanderhoof, *The Old Mechanics' Bell Tower*, 1882, Picture Collection, New York Public Library. Put up in New York's shipbuilding district in the 1830s, this iconic bell represented the efforts of laborers to own their time.

the *Liberty Bell*. In various iterations, the bell was imagined as a chaotic force, one that would reorganize society as it spread freedom.[18] Although its role as a timekeeper receded, the Liberty Bell was unmatched in importance as a repository of national aspiration. The conflict over slavery, expansion, and the meaning of freedom catapulted the Liberty Bell to iconic status found imprinted on stamps, currency, or war bonds. It became America's preeminent symbol, summing up national aspirations and ideals.[19]

The Old State House's importance as a source of clock time for Philadelphia residents solidified in these decades. In 1845, the clock acquired new hands. In 1852, its wooden dials were replaced with ones of solid ground glass whose diameters extended seven and a half feet. The constant indication of the time on the state house's enlarged and brightened clock dials attested to its imposing visual presence. A few years later, in 1857, the city introduced an alarm system rigged to ring all the bells and gongs in the city in the event of a fire. The system was also used to send time signals from the state house clock to the city's bells and fire-alarm signal boxes, which together rang at noon every day.[20] The state house clock was at the center of the network of aural and visual time signals. The sound of time ringing forth every day knit various parts of the city together around state house time.

Beginning in the 1840s, railroads began to gain outsized authority for the time, largely because it was their time that communities needed to keep if they wished to post letters, ship commodities, or travel. Railroad timetables, reliant on time standards for their very practicability, proliferated. Their schedules ensured that their time would be inscribed atop local hierarchies. Not only did railroads acquire clocks for their stations and watches for their conductors, they also demanded punctuality from their passengers.[21] Initially a railroad's standard of time corresponded to the local time of its headquarters and was shown on the railroad superintendent's office clock. For instance, the officers of the first interregional railroad in the United States, the Western Railroad, mandated in 1843 that "the clock at the upper depot in Worcester [Massachusetts] shall be taken to be the standard time." Soon the method of using an agreed-upon meridian, rather than a designated clock, for a time standard gained some powerful adherents. Meridian time eliminated the imprecision of using a clock as a standard; it was a scientifically objective measure of time. It could be used to synchronize clocks set at great distances from each other. The telegraph (introduced in 1844) simplified the dissemination of meridian time and enhanced the potential to synchronize clocks. A region connected telegraphically could maintain a single standard of time, and that standard could be an agreed-upon meridian. Great Britain introduced a national standard time in 1848, using Greenwich as its meridian and telegraphy to synchronize its clocks. The next year, New England railroads adopted "two minutes after the true time at Boston"—that of a meridian just west of Boston—as a "single standard time" for their operations.[22]

Though demands for temporal accuracy were accumulating before the introduction of the telegraph in 1844, they intensified afterward, spurring the establishment of astronomical observatories throughout the United States. Nearly half of the observatories built before 1855 went up after the first telegraphic relay between Washington, DC, and Baltimore, in which the fifth message asked, "What is your time?" As the historian Ian Bartky comments, "From that moment on, Americans were captivated by the near-instantaneous transmission of time to distant places."[23] An observatory was built in Washington, DC, with funding from an 1842 act authorizing the navy to build "a depot of charts and instruments." Colleges, associations, and individuals also built astronomical observatories, two dozen across the country, including one stocked with instruments made in Germany situated behind the Philadelphia High School.[24]As one booster explained, an astronomical observatory "furnishes an accurate determination of the time, which is a matter of importance to almost every citizen."[25]

After 1851, owing to the innovative work of the Boston instruments maker and astronomer William C. Bond, time signals could be telegraphed directly from an observatory's clock, enabling the development of time services. Like other fine instrument makers in American port cities, Bond's shop rated, repaired, and provided on "loan chronometers for long and short voyages."[26] Rating required access to accurate time so as to document a timepiece's tendency to run fast or slow; with this information, it could be regulated while at sea or when otherwise far from astronomical instruments. Bond operated his public time service in cooperation with the Harvard College Observatory. New England railroads, local jewelers, and the city of Boston subscribed to the service. During the 1850s, four other American observatories also operated fledgling time services in which they provided the local time to watchmakers, jewelers, and clockmakers.[27] Watch- and clockmakers engaged in the sale, rental, and rating of chronometers received time signals in their shops over dedicated telegraph lines from local observatories. Previously they had determined the "true time" using astronomical instruments and charts.[28] These time services perfected a way to maintain a standard of time: observatories would determine the "true time" of a particular meridian; telegraphs would disseminate it to time specialists; clocks, bells, and time balls would radiate it; and timetables would turn it into the coin of the realm. As Carlene Stephens puts it, "More than any other technical factor, the development of time-distribution technology systems based on telegraphy prepared the way for a uniform national standard time."[29] It was several decades, however, before this new way to distribute the time was widespread enough to precipitate such a significant change in American timekeeping practices.

The US Navy erected its first time ball in 1845, emulating the one that went up in Greenwich, England, in 1833. Located on the banks of the Potomac River, the navy time ball served less as a navigational tool and more as a symbol and monument.[30] Just as a bell only indicated the time when it was ringing, and otherwise conveyed no temporal information, a time ball only indicated the time at discrete moments, that is, during its drop. Typically perched atop the highest point in the central part of a city, usually a tower, these globes with metal ribs and canvas covers of various colors were rigged to an electric pulse, which caused them to drop at noon. First built in England in 1829 as devices meant to help in the rating of ships' chronometers, time balls never numbered more than a hundred worldwide.[31] They were intended to convey through visible means the exact time: after 1844, telegraphic wires connected astronomical observatories' accurate clocks (known as astronomical regulators) with the apparatus meant to release a time ball. The time ball drop ostensibly allowed the clockwatching public the opportunity to set its watches and clocks. But a time ball on the banks of the Potomac served practically no one: few oceangoing ships navigated the shallow river, most Washingtonians clustered far from the river's edge, and the ball itself, released by hand rather than telegraph time circuit, could not convey accurate time.

What is more, few had the need for accurate time in the Washington of the 1840s and 1850s. The newly made "large Town Clock, Made for the City of Washington" at a cost of nearly $3,000 dollars was sold on the auction block in 1847, rather than installed.[32] To be sure, in the Senate's chambers ticked an eight-day clock made and installed by Simon Willard shortly after the turn of the century. In the Capitol's Statuary Hall presided a marble Clio, the muse of history, in a winged chariot atop a globe with a clock beneath, sculpted by Carlo Franzoni in 1819 and with clockworks by Simon Willard. In the Supreme Court's chambers yet another Simon Willard clock took pride of place, this one installed in 1837.[33] But the clocks only loosely governed the operation of Congress or the Supreme Court.

Indeed, by the 1850s, one Senate custom had an attendant turn the hands of its clock back, so as to see appropriations bills approved in a timely manner or to avoid going into extra sessions. Senate employee Isaac Bassett, who had charge of this duty, recalled the practice in 1895, near the end of his sixty-two years of service:

On a great many occasions where I have put the clock back a great many of the senators would call me to their seats and question me as to what authority I acted upon. I always answered with great dignity, "By order of the president of the Senate of the United States." They would say, "By what authority does he order you to do that?" I told them that it has always been the custom at the last hour of the session when they were waiting for important appropriations bills to come from the House to turn the clock back.[34]

These senatorial challenges were put to Bassett in the 1880s and 1890s, after the standardization of time, when clocks were understood as immutable conveyances of the time. But in the middle decades of the century, matters

Harry O. Hall, *Isaac Bassett Turning Back Hands of Senate Clock at Close of a Session of Congress*, 1892, LC-USZ62-101776, Prints and Photographs Division, Library of Congress. Captain Bassett, employee of the US Senate for over sixty years, is depicted adjusting the time to avoid the appearance of overlong and inefficient Senate sessions. Basset's manipulation of the clock was understood in the antebellum period, but increasingly came under scrutiny during the public clock era because clock time was gaining status as an irrevocable scientific fact, not a convention to which many subscribed and some could manipulate.

stood differently. As Bassett explained, "in olden times when I turned the hands of the clock back the senators all understood it but in these latter days all seems different."[35] Time served the state; in its instability, it was malleable. The clocks found in the US government's ruling chambers and public spaces conveyed authority, legitimacy, and even a grandiose vision in which historic and mechanical time were together triumphant. During the middle decades of the nineteenth century, the few timekeepers the national government installed aspired not to transmit the time so much as to announce authority and mastery.

In effect, the railroads' usage of a standard of time, whether observatory time or local time, created an incentive for others to follow. But for decades railroad depots were on the edges of town, where any public clock would not necessarily be widely visible. Railroads could adopt a local time standard as their own, or they could look to local authorities to disseminate their time. In the latter instance, they depended particularly on jewelers, whose long-lived claim to temporal authority was sustained by their time-telling expertise, inventories of clocks and watches, and prominent display of well-regulated clocks known as regulators. As the superintendent of the US Naval Observatory explained to one US congressman, "jewelers are considered the repositories of correct local time."[36] Calls were made again and again for the "true time," that is, for clocks to be set to mean solar time, but the real problem was that most clockworks were unreliable, and what is more, the means to synchronize clocks were not yet widely accessible.

As one New York paper observed in 1856, "[If the city] has any one need more urgent than another, the true time is that very need." What the piece described, however, were not the outrages committed for lack of access to the "true time," but the "chronic incongruity among the great clocks." The article pointed to the city hall's clock dials, the dial on the façade of jeweler and clock manufacturer Sperry and Company's building, and the clockworks that rang the bells of Trinity Church. It conjured a New Yorker following city hall time trying to do business with a friend on "the exchange" regulated by Hammond, a cashier with "Benedict in his pocket," and a notary carrying Tiffany.[37] Here the author knowingly invoked the city's three most prominent jewelers: the watch importer Samuel Hammond, the jeweler Samuel Benedict who claimed to be the keeper of New York's city time, and Tiffany's famous Atlas hoisting a clock rather than a globe.[38] In crying out for "a standard of time," as the report was titled, it identified six possible time standards that city residents followed, four of which jewelers set.

Shortly after the incongruity of the city's clocks was noted in the press,

S. W. BENEDICT,

WATCH MAKER.

Nº 5 Wall Street,

NEAR TRINITY CHURCH

Has removed from the Merchants Exchange to No. 5 Wall Street, where he has opened an entire New Stock of Watch and hopes to continue the reputation he has had for the last Fifteen Years, for selling Fine Watches. No pains or expen has been spared on his New Regulator, and the public can rest assured of its keeping the correct Time. All of the W. Street Expresses, and most of the Steam Boats and Rail Roads start by it.

He has made a permanent arrangement with Mr. Cottier, who has been Foreman for him for the last Three Years, ar great care will be given to the Repairing of Fine Watches.

T. F. Cooper, is supplying him with his best Chronometer and Duplex Watches, which will be sold as low as if purchase. of him in London; He has also the Anchor Escapement Watch, a very handsome pattern for Ladies, together with Roskell Tobias, and Beesly's Lever Watches.

Spoons and Forks warranted Sterling Silver, French Mantel and office Clocks, Jewelry, &c.

Trade card for watchmaker S. W. Benedict, ca. 1840, New-York Historical Society. Benedict claimed to set the time for New York City. This elaborate flier justifies the jeweler's monopoly over city time: his shop's proximity to Trinity Church's chimes and clock, deep inventory of imported watches, and starting point for omnibuses and carriages. Add to that a precise timekeeper and an accurate method of determining the time, and a jeweler like Benedict could be an authority for city time at midcentury.

Advertisement for H. Sperry and Company's "Manufactory of Tower Clocks," ca. 1858, New-York Historical Society. Sperry's was just one of the many time standards a New Yorker might have followed through the 1860s.

TIFFANY & CO.

LATE TIFFANY, YOUNG & ELLIS,

THE LARGEST STOCK IN THIS COUNTRY OF

Fine Jewelry, Diamonds, and other Precious Stones,

Charles Frodsham's, Cooper's, Jules Jurgensen's and Patek Philippe & Co's Watches.

SILVER WARE,

Table Cutlery, Bronzes, Clocks, Rich Porcelain, Fans, Desks, Dressing Cases, Work Boxes, &c., &c.

550 BROADWAY, N. Y.

House in Paris, TIFFANY, REED & CO., Rue Richelieu, No. 79.

Every article is marked in plain figures at the lowest price for which it will be sold. An inspection of their stock incurs no obligation to purchase. Equal attention will be shown visitors whether their calls are for business or pleasure.

Advertisement for Tiffany and Company, 1856, New-York Historical Society. Tiffany's Atlas carried a clock on his shoulders, rather than the earth, signifying a shift from space to time as the most important factor in dominion.

representatives of the US Coastal Survey and of Albany's Dudley Observatory "sent a message to the Mayor expressing a desire to supply New York with time." They promised its accuracy "to the tenth of a second." Proposed were a time ball and the transmission of the "true time" through telegraphic circuit to "any clocks which the City Government may select." The need for such a service was evident. As the *New York Times* reported, "The great commercial port of New York not merely has no precise signal for the regulation of chronometers, but is dependent upon private resources for a knowledge of the time by which the clocks are to be regulated."[39] Four years later, a time ball was installed atop the city's US Customs House, regulated by telegraphic signals sent from Albany's recently built Dudley Observatory. Its unreliable operation soon became a source of satirical humor in the New York and Brooklyn press.[40] Other clocks in New York and Brooklyn disseminated the time, but there was still no single source for city time. Indeed, there was not a single standard of time anywhere in the United States except Boston and its environs.

Returning to the 1830s, we now consider the growing extent of personal timekeepers throughout the country. The flow of household clocks was so great, and that of watches so feeble, that two, three, four, and even five times more clocks than watches could be found in almost any American community.[41] In port towns and farming communities, jewelers occasionally rented out watches, loaning chronometers to ship captains and everyday watches to farmers heading to the city.[42] By the first half of the decade, several dozen Connecticut clock factories were producing an estimated 38,000 clocks a year. In 1836, their production reached 80,000. It appeared as though, as one manufacturer put it, these clocks "multiplied like the leaves of the forest." A network of Yankee peddlers sold these timekeepers to households in rural areas, small towns, and along the frontier. Although Connecticut clocks dominated the trade, several dozen German peddlers traveled throughout the United States selling clocks made in the Black Forest region.[43] Occasionally clock peddlers also sold imported watches, but the vast majority of their trade was in clocks.[44] After recovery from the financial panic of 1837, which shut down most clock production for a few years, manufacturers turned their attention to brass clocks owing to improvements in the process of rolling brass. Brass clocks were far more reliable timekeepers than ones with wooden movements; through the 1840s, domestic and foreign demand for these clocks soared. By midcentury, more clocks were for sale in the United States than ever before. European clocks were available in larger cities, and in smaller towns a local merchant might have one or two in his inventory.

American-made clocks could be found everywhere. Chauncey Jerome's two Connecticut factories, for instance, produced 280,000 brass clocks each year, which were sold for as little as $1.50 each.[45]

In the 1850s, watch importers in the United States picked up the pace of their trade with Europe, importing many more watch movements and parts for assembly into watch movements than previously.[46] During these years, European watchmakers intensified their solicitation of business with American jewelers: one of the founders of the Swiss firm Patek Philippe toured the United States in 1854 and 1855; he set up exclusive partnerships with the most esteemed retailers of each of the largest cities. At the high end of the market, European origins connoted value and prestige, but at the lower end they could be overlooked or obscured. Some American firms engraved their signatures and establishments' place names onto assembled watch movements made of European parts. New York City's Samuel Hammond and Company occupied a middle ground: it advertised its importing business without reference to the names of the London watchmakers whose wares it sold, but instead emphasized that the pocket chronometers and lever watches it offered were "manufactured expressly for us in London."[47] Working on the wholesale side of the business, watch importers also sold small batches of watches to jewelers engaged in retail business. Reflecting the rising volume of business, specialized signs appeared: an 1855 flyer for "Emblematic Signs (Carved or plain figured)" suggested that "Eagles can have a Watch suspended [*sic*] from the beak, for Jewelers."[48]

Not only did the market for watch imports grow during the 1850s, demand for secondhand watches was also strong. As the historian Wendy Woloson explains, pocket watches "circulated promiscuously in retail, resale, and underground markets." She characterizes the pocket watch as "a reliable form of mobile capital," suggesting that its value was likely greater as collateral for loans and as a status symbol than as a timekeeper. People pawned watches at great rates; 14 percent of 27,000 pawns recorded in one antebellum ledger were for watches. During the second half of the nineteenth century, the pawn business relied on watches for a third to nearly a half of its trade.[49] Many of these pawns were not redeemed; instead the watches entered the secondhand market.

Despite rising imports and a flush secondhand market, the distribution of watches among the population remained constant and fairly low through the 1850s. This is far longer than might be conjectured given the increasing pressures toward time discipline and time orientation. In fact, no larger a proportion of the Massachusetts population in 1870 owned watches than did that of Long Island and Connecticut in 1800, or the population of Delaware or the Hudson Valley in 1820, or that of South Carolina in 1850. What is more,

Trade card for watch dealer William B. Eltonhead, ca. 1855, tinted lithograph by John L. Magee, courtesy of the Library Company of Philadelphia. In the 1850s at the establishment of Philadelphia jeweler William B. Eltonhead, a lively trade gathered around elegant picture windows captioned "Watches and Jewelry" and beneath a double-dial clock affixed to brackets between the building's second and third floors.

there is no correlation between the ownership of a watch and accumulated wealth or occupational status during this period. The poorest and least skilled were as likely to own a pocket watch as the wealthiest and most skilled.[50] Designed for the trouser-and-jacket-wearing classes, the pocket watch and chain were nevertheless part of the furnishings of many men who wore smocks and aprons to work.[51] That manual laborers would have been hard pressed to find a place for their pocket watches in their workday clothes simply highlights the watch's lack of utility in most nineteenth-century workplaces. But in public places a different story unfolded. As one historian concludes: "A watch was a very public item of clothing."[52]

And indeed it was, as one story from the 1840 campaign trail reveals. In the midst of a debate preceding the 1840 presidential election, a Democrat opponent of the young Abraham Lincoln who was stumping for the Whig nominee, William Henry Harrison, began "denouncing the aristocratic principles of the Whigs." The campaign turned on the contrast between log cabins and mansions, "the people" and aristocrats. As the Democrat went on about the Whig's love of finery, Lincoln, according to a historical account, leaned down and with a slight tug "pulled on his opponent's vest, which popped open to reveal a ruffled shirt, gold watch, and chain." The crowd roared with laughter. Clearly it was the Democrat with his gold watch and chain who lived by aristocratic principles. By the time he served as president, Lincoln himself owned at least two gold watches. Even then the necessity of a gold watches was as doubtful as that of a pianoforte; they were taxed at rates between one and two dollars per year in the federal income tax levied between 1862 and 1872. Silver watches were exempt from such taxation.[53] Altogether pocket watches as luxury items bore the weight of more than just a tinge of old-world presumption.

So, taking these circumstances into consideration, it was with somewhat poor sales prospects that in 1849 inventor Aaron Dennison convinced his father-in-law Samuel Curtis and Boston clockmaker Edward Howard to found a company that made watches using interchangeable, machine-made parts. Through most of the first half of the nineteenth century, relatively weak demand for watches, expensive entry costs into the industry, a lack of skilled watchmakers, an abundance of secondhand watches, and the steady supply of European watches had together deterred Americans from starting up watch manufactories.[54] Over the next several years, Dennison, Howard, and Curtis built a variety of machines to make watch parts and named their enterprise the Boston Watch Company. They moved their factory to Waltham, a suburb of Boston, where a hundred employees made between six and ten watch movements a day, which sold for forty dollars each.[55]

The Boston Watch Company's output in the 1850s was considered sizable. It was only because the company's founders had extensive networks among jewelers that they were able to sell their watch movements, along with fitted silver cases made to go with them. Jewelers assembled finished watches from the variety of standard-sized movements and casings. The Boston Watch Company's factory also produced spare parts for its watches, reducing the need for custom-made replacement parts. In all, these attributes increased enthusiasm for American watches. Nevertheless, continued preference for European watches, a preponderance of secondhand watches, and a business recession in 1856 stymied sales of Boston watches. Demand for machine-made American watches was further depressed because they were priced well above medium- and low-grade Swiss and English watches. By 1857, the Boston Watch Company was bankrupt.[56]

The Boston Watch Company's factory and its contents were auctioned off on a rainy day in 1857 to New York watch wholesalers D. F. Appleton and Royal Robbins and a firm of Philadelphia gold watchcase makers, Tracy and Baker. The auction was fierce, suggesting not only that the founder Dennison and his partners "succeeded technically, even if they fizzled financially," but also that the business community anticipated that a market for American watches would take shape. After an interim period in which few watches were made, Royal Robbins emerged in 1859 with control over 85 percent of the Boston Watch Company, whose name he changed to the American Watch Company.[57] Robbins understood that "Americanness" enhanced the value of his company's timepieces; there was a growing market for *American* watches.

Several important developments in the watch industry ensued. First, during the last two years of the 1850s, Edward Howard, who had forfeited his interest in the Boston Watch Company, formed E. Howard and Company, which pioneered several significant innovations in the design and manufacture of watch movements, while also building a business in public clocks.[58] Second, Royal Robbins's reorganized American Watch Company acquired the machines, materials, and workforce of a rival watch company, which he set up in a separate workshop at Waltham to make high-grade watches competitive with E. Howard's products and with foreign imports.[59] Third, in 1859, a jeweler named James Boss patented "spinning up," an inexpensive technique for making watchcases. At this time, many more watchcases than movements were made in the United States.[60] By the early 1860s, the mass manufacture of a variety of grades of watch movements and cases was possible. The only thing missing was a market large enough to absorb them.

Just as industrial capitalism was gaining momentum in the 1850s, so was the reformulation of temporal mandates and practices. The Civil War did not introduce a new consciousness of time, but it accelerated the halting and slow embrace of clock time that had begun a century earlier.[61] Pressures toward the ownership of time had been accumulating for decades. In the 1820s and 1830s, clocks had become useful tools for measuring productivity, achieving efficiencies of scale, and synchronizing social activity.[62] Since the mid-1840s, railway and steamship timetables encouraged shipping customers and passengers to temporally synchronize and coordinate their use of the transportation network. Telegraphs sped market transactions, placing an even higher premium on control over time. Even in the home, mechanical time gained force; schedules were valorized as part of an orderly household, and clocks were used to time cooking, baking, washing, and other household labor. The usefulness of clocks and clock time was everywhere becoming evident. New techniques and practices dependent on the telegraph simplified the coordination of clocks. By 1860, the authority of clocks and watches seemed more than chimerical, more than a figment of the Gothic imagination as brought to life in the fiction of Edgar Allan Poe.

The inability to marshal time, owing largely to the impossibility of synchronizing timepieces, characterized the 1860s, especially the war years. The absence of standard time, according to one historian, "proved disastrous for both the Union and Confederate armies."[63] Most orders stipulated times for movements, but since there was no single standard of time, nor a master clock providing a uniform time, it was hard to follow these orders with the necessary precision. There were few clocks in camps, but many pocket watches, although how soldiers cared for them is not entirely clear. They could only be wound with keys; daily they lost seconds or even minutes; and changes in the temperature or humidity level often destabilized them. Soldiers set their watches by their officers' watches, or by local time as made apparent by a nearby town's most prominent clock tower or bells, or by makeshift sundials set up in camp to determine local noon. Field officers tried to rely on the telegraph to regulate their watches and synchronize them with command officers' timepieces, but more often than not this tactic failed.[64]

It is no wonder that temporal confusion and mistakes characterized some of the more humorous as well as some of the more tragic aspects of the war. One bugler serving in a Pennsylvania regiment near the end of the war entertained his family in a letter with just such a story. He woke up in camp, looked at his watch, and seeing that it was "6 O'clock," he "picked up the bugle and played the Reveille." Within five minutes, according to his account, "the drummers and buglers in our whole Brigade were out playing the Reveille." Soon the young bugler received a note from regimental headquarters; when he presented himself to the colonel, he was asked "what in the Devil [he] was about in blowing the bugle so early" and was ordered to look at his watch, which still showed "6 O'clock." The soldier "knew then that the watch had stopped." To make the humiliation worse, the hapless bugler reported to his father, the colonel then "drew out his watch and says 'here I will show you what time it is, fifteen minutes past one.'"[65]

Perhaps after this middle-of-the-night incident, the bugler sent his watch home for repair, as did many soldiers. For instance, Sylvester McElheney, a private in the Union army, sent his watch home to "getit fics" just a week before he died of battlefield wounds.[66] Franklin Rosenbery, another soldier fighting for the Union, carried on an extensive correspondence about his broken watch, pleading with his father to take care of the repair: "Tend to the getting of the watch fixed as soon as you can as I need it very much." He wrote repeatedly for his watch, several months later confessing: "I am very anxious about my watch. I would like to have it as soon as possible as I need it very much." Injured six months later, Rosenbery wrote to his family from a Philadelphia hospital, asking for his watch again: "I would like it if you would express my watch to this hospital."[67] Union soldier George Thomas sent his watch home, explaining: "It has got out of order and woant [*sic*] run This I much regret for I have got so accustomed to have the time that I will be very much at a loss with out it But there is no possibility of getting it re-paired here And I can't think of carrying a dumb watch so I send it home."[68] These entreaties draw attention to how important well-running watches were for soldiers. If the watches were simply status symbols and talismans, then they could have been kept in closed cases and on chains, but for a host of reasons—status included—knowing the time, if just an estimate, mattered, even to hospitalized men.

More important to the experience of time during the war than broken watches was the inability to expediently and effectively coordinate move-ments and complete tasks. In the battle of Bull Run, massive delays caused by poor planning, bad weather, and lack of seasoning prevented Union troops from moving as quickly as possible, all of which were factors in their rout. The Confederacy managed to win the battle because, as historian Cheryl Wells explains, "coordination, temporal precision, and unity of action were less important" to forces on the defensive. Confederate General Beauregard's watch broke during the war's first major land battle, and his army was not able to follow through on various orders stipulating the time for action. Even so, the Confederacy's defensive position rendered following clock time less crucial than it was for the attacking Union troops, who ended up retreating disastrously.[69] On the defensive at Gettysburg, Union troops prevailed de-spite delays for both sides. As one Confederate officer succinctly explained: "From the first to the last, each move was made just an hour too late and this is the story of Gettysburg."[70]

Battles, like other cataclysmic events, were understood to have happened in calendar and clock time, and occasionally in religious time. In military correspondence, newspaper reports, and memoirs, clock time nearly always denoted the beginning, duration, and end of battles.[71] Battles unfolding out of step with plans cued to clock time were, counterintuitively, a source of an enhanced and widespread devotion to mechanical timekeeping and time-keepers. The potential of coordinated, timed action was recognized. It would come to influence strategies for waging war in the twentieth century, but in the immediate aftermath of the Civil War this potential, in concert with an expanded railroad and telegraph network, hastened the pace of day-to-day life. Such quickening made the usefulness of coordinated public clocks seem self-evident, and the necessity of personal timekeepers obvious. It is no surprise that demand for watches continued to rise after Civil War soldiers mustered out. The machine manufacture of quality watches in the United States and their martial associations transformed these beguiling instruments from signs of European prestige into symbols of American ingenuity.

CHAPTER 3

Republican Heirlooms, Instruments of Modern Time Discipline

Pocket Watches during and after the Civil War

The American watch has eminent claims as the true republican heir-loom—a triumph of industry in an age of industry, it symbolizes the progress and dignity of labor; a product of American enterprise, it is associated with the sentiment of patriotism; moderate in cost, it is accessible to the body of the people, and, thoroughly made, it is prepared for a lengthened future. . . . When a hundred years have rolled away, and the continent is reclaimed to civilization, and telegraphs enclose the globe like a net, a white-haired man shall say: "My son, when I pass away I shall leave you this watch."

— "The Watch as a Growth of Industry," *Appleton's Journal* (1870)

Many soldiers returned from the Civil War with recently acquired pocket watches. They also bore a newfound interest in mechanical time and timekeeping. One of these soldiers was David Edwards Hoxie, a twenty-three-year-old native of Roberts Meadow, a cluster of farms about five miles north of Northampton, Massachusetts. Receiving a medical discharge from the Union Army in March of 1863, he informally apprenticed himself to a local Northampton jeweler and watch repairman, D. Foster Davison.[1] Under the jeweler's instruction, the Civil War veteran learned to repair watches; by 1871, he had parlayed that skill and some capital into a partnership with Davison, which according to credit reports did "a small safe business" in watch repairs. A few years later, Hoxie opened a small watch repair business of his own, which he kept running until the mid-1880s.[2] Hoxie's business was one of hundreds, if not thousands, devoted to watch repairs and sales that sprang up around the United States during the 1860s and 1870s.[3] Like others who worked with valuable objects, Hoxie kept meticulous records of each repair job in a watch register. His was a large, leather-bound book with multiple columns to record the date and client's name; the watch's signature, serial number, materials, and other identifying characteristics; and the repair work and its cost. With such detailed information, what Hoxie called his "watch book," like other registers, is a rich source of information about the circulation, meaning, and use of watches.[4] Nearly one out of five of the timepieces Hoxie repaired were Waltham watches, which during the Civil War had gained fame for being machine and American made.[5] Just as for Hoxie and the American Watch Company of Waltham, the Civil War heightened the importance of mechanical time and timekeepers in the United States.

So too did appeals to patriotism during and after the war. The *Appleton Journal*'s 1870 feature article on the machine manufacture of watches cast a Waltham watch, kept in an old man's family for "more than a century," as a repository of patriotic sentiment and personal memory. It had been a grandfather's "inseparable companion" through the Civil War; it had "counted out the precious minutes for two generations"; it was "a triumph of industry." The article's white-haired narrator anticipated that the Waltham watch would mark the hour of his death.[6] Laden with the associations of war, death, family, and country, this watch could be nothing other than an heirloom. What made the Waltham watch "republican" rather than "monarchical" was that it, and others like it, fostered modern time discipline. A watch enabled its owner to inwardly note and track the time. It engendered independence. "Once the luxury of the rich, it was now the necessity of all": the pocket watch was an indispensable part of the self-made American's tool kit.[7] It was a vehicle of self-realization, rather than of privilege and inheritance. As a symbol of industry, democracy, and patriotism, then, the American watch staked "eminent claims as the true republican heirloom."

If not American made, a pocket watch could still lay claim to national significance had it been owned by one of the nation's founding mothers or fathers. The trustees of Washington's Headquarters, a historic site in Newburg, New York, purchased in 1879 the watch George Washington gave Martha Custis shortly before they married. At the time it was felt that the English watch with Martha's full name overlaying the numerals was "a most valuable and interesting relic."[8] One of Abraham Lincoln's watches also became a storied item upon which, literally, words were exchanged with regard to the

Occupational portrait of a watchmaker, ca. 1840–60, daguerreotype, LC-USZC4-4072, Prints and Photographs Division, Library of Congress.

Civil War. In April 1861, a watchmaker secretly inscribed Lincoln's watch movement with his own name, the date, and the comment "Fort Sumpter [*sic*] was attacked by the rebels on the above date." The inscription ends "thank God we have a government." Years later, the repairman recalled that when he worked on Lincoln's watch he was the only Union man in M. W. Galt's Pennsylvania Avenue watch shop. Another engraving on the movement, added who knows when, taunted with the words "Jeff Davis." What the *New York Times* has called a "timeless Lincoln memento," this watch, acquired by the Smithsonian Institution in the 1950s, is perhaps the ultimate "republican heirloom."[9] As American-made watches came of age during the middle decades of the nineteenth century, they also accrued an array of patriotic and sentimental associations that the phrase "republican heirloom" aptly captures.

Patriotism added value to watches in other ways. The marketing pitches built on nationalism had the added benefit of keeping pressure on the US Congress to renew high tariff rates on imported watches.[10] Patriotic designs and sentiments on watch casings underscored the association between watch ownership and nationalism. As the celebrations of 1876 approached, manufacturers gave their watches patriotic signatures, such as the Waltham signature "Centennial." Some trade cards used patriotic imagery, particularly the eagle and the American flag. An extant Brooklyn Watch Company case from the 1870s shows a sailor holding an American flag, with a figure of Liberty placing a wreath on the sailor's head. Shortly after 1900, the Suffolk Watch Company marketed "Betsy Ross" watches, with the pitch: "Sentimental and historical in name—Reliable as timekeepers." Around the same time, the United States Watch Company sold a high-end watch with the signature "The President."[11] The association of a man's watch with the presidency continues to sell watches to this day. The Presidential Watch Company sells the same watch that Barack Obama wore on the night of the 2008 election and the day of his 2009 inauguration.[12]

In this chapter, I look closely at watches during the Civil War and then through the end of the nineteenth century. After the war, local authorities took more responsibility for the distribution of accurate time than they had before; at the same time, the owners of personal timepieces expressed more interest in setting them accurately. Pocket watches personalized and individuated the public time of tower clocks and hour's bells; they fed a sense of ownership of the time, while reinforcing the ties between the self and society. These devices provided a bridge from a world in which natural time held sway to one in which clock time ruled. Demand for watches ballooned. Concurrently, rituals associated with watches, such as presenting them to victorious athletes, community heroes, retiring employees, and the betrothed deepened and spread during the Gilded Age. I explore the considerable attraction watches held for Americans of all stations. In doing so, I suggest that pocket watches helped acculturate Americans to mechanical time. They heightened sensitivity to the very passage of time, while providing a tool with which to achieve greater degrees of time discipline. Ownership of pocket watches accustomed Americans to carrying the time with them, made them eager to

Trade card for J. H. Johnston and Robinson, ca. 1860–70, Advertising Ephemera Collection, Baker Library Historical Collections, Harvard Business School. During the Civil War, New York agents for Waltham's American Watch Company issued a trade card with an American eagle perched atop a pocket watch, reminiscent of the Liberty Bell's cap.

consult public clocks for accurate time, and readied them for modern time discipline, a condition in which clocks provided a common organizational notation.

War accelerated the transformation of Americans, particularly soldiers, into clockwatchers. During the US–Mexican War (1846–48), most officers and some soldiers carried watches, but it was during the US Civil War (1861–65) that watches spread through the ranks.[13] Mastering mechanical time was preeminent in the plans, hopes, and efforts of the Union and Confederacy. As the historian Cheryl Wells shows, both the Union and Confederate leadership routinized camps, hospitals, and prisons according to the clock, while also attempting to implement battle plans based on clock time. These various experiences with clock-based discipline imprinted themselves on soldiers. Their memories of the war often included pinpointing the time and duration of marches, attacks, battles, retreats, and deaths.[14]

Neither the Confederacy nor the Union was able to follow timetables, schedules, or orders with temporal precision. The inability to execute timed movements, however, did not lead to a repudiation of the clock. As in armed conflicts that preceded the Civil War, timing was an important aspect of planning for and waging campaigns. Evidence about the experience of time in the various sites of the war suggests that time and task orientation together served as guides and motivation throughout the conflict.[15] Time discipline and time obedience were also intertwined. Soldiers carrying watches wanted to be able to anticipate (and later to note) the hour of rising, of drills, of movement, of battle: here time discipline was at work. Yet rousing troops to action still required young boys to play the reveille on their bugles and commanders to make aural and visual commands; that is, it called for time obedience. Together the mixture of orientations toward tasks, times, and authority created an intricate set of mandates and compulsions in the theater of war as much as in everyday life.

If imperfect instruments, timepieces were nevertheless considered vital to the conduct of the war from the perspective of the draftee and the West Point general alike. The very design of the Union's uniforms suggests that officers and enlisted men were expected to carry a watch. An extant single-breasted wool jacket worn by a general, for example, sports on its right side a watch pocket. Pocket watches were typically carried in vests, but since not all soldiers wore vests, a watch pocket lined with white cotton might have been sewn to the right of the fly of soldiers' trousers, as is the case with three pairs held by the New-York Historical Society. Pocket watches were not simply ornaments; they were considered necessities, like shoes, rifles, and canteens.[16]

The engraved sentiments and scenes on the casings added to the watches' utility, for they reminded soldiers of their commission and duty. A shield engraved onto the casing of one watch carried by a career officer in the Union army, for instance, illustrates this point.[17] Another pocket watch, made in the late 1850s, was inscribed with the name D. G. Farragut, the date 1861, and the command "To Preserve the Union."[18] The watches soldiers carried during the war gained inestimable value: more than forty years after the war, a lieutenant in the Union army left instructions that the English-made gold pocket watch that his mother gave him in 1860 and that he carried throughout his three years service in the war be kept "in the family."[19] In 1863, a lieutenant in the Union army, Charles M. Walton, had a carte de visite made of himself in uniform gazing at an open pocket watch. One of these cards, inscribed to his "dear friend O.W. with love," was presented to the future Supreme Court justice Oliver Wendell Holmes. The alchemy of photograph and mechanical timepiece produced a powerful elixir, in which attention is drawn to the inevitable passage of time as well as its momentary arrest in both the photographic image and watch dial.

There was a large and substantial market for pocket watches among soldiers on both sides of the conflict. "Roving merchants," as Carlene Stephens

To my dear friend O.W. with love from C.M. Walton

Portrait of Charles M. Walton, April 1863, photograph by J. W. Black mounted as carte de visite, Oliver Wendell Holmes, Papers, Photographs, and Other Images, ca. 1856–1934, courtesy of Historical and Special Collections, Harvard Law School Library.

explains, "sold thousands of cheap watches to eager customers in wartime encampments."[20] But not all soldiers availed themselves of the opportunity to buy a watch from a peddler. In their first letters home, many soldiers requested a watch. One private in the Union army asked his father to get him "a cheap watch . . . not more expensive than eight dollars." He explained, "There are plenty of watches here [in camp] but I do not know any thing about a watch & people's conscience dont [*sic*] trouble them about the soldiers."[21] Soldiers also bestowed watches on loved ones, especially to fiancées and wives.[22] Sometimes friends and families gave watches as gifts to soldiers; for instance, a soldier serving in the Ninth Pennsylvania Reserve Infantry received a watch and a revolver from his brother just months after the war began.[23] Soldiers who preferred not to rely on friends and family ordered watches from retail jewelers by mail. New York City's mail-order firm Hubbard Brothers offered to send watches, as well as gold pens, to "any part of the Loyal States, with bill payable to the Express man." Apologetically their catalog, which was smaller than a three-by-five index card, demanded "CASH IN ADVANCE" from "soldiers and all others in Disloyal States."[24]

An active trade in watches among soldiers in the camps of the Union and Confederacy also sped their circulation. The journal of an infantryman fighting for the Union, for instance, intersperses details about watch trades in between notations concerning the duration of drills, the arrival times of officers, and the departure times of troops. Over the course of two months, the soldier "bought a watch," went "up to the new regiment to trade watches," "sold watch chain," "traded watches," traded his watch for a fiddle, and bought a watch for five dollars (which he sold a few days later for twelve dollars).[25] Another young man serving the Union as a commissary officer, the son of an Ohio jeweler, wrote home: "If you think watches are the best investment, get some and keep them at home till I send for them. . . . Get me An American the balance nice fancy anchor levers that I can sell from 18 to 22 dollars."[26] Another officer also earned extra money by speculating on watches. He wrote to his family, "I am glad the watches are a coming as I think I can dispose of them to good advantage."[27]

In addition to trading watches, soldiers also confiscated them from civilians, prisoners of war, and the dead. During the Confederacy's drive toward Gettysburg, troops passing through Chambersburg not only took "hats & boots off the men," and "$50 off Dr. Sneck" but also "his gold watch valued very highly."[28] In the immediate aftermath of the war, it was rumored that "Yankee soldiers robbed [African Americans] of money, watches, and all valuables."[29] A memoir of a Union prisoner in Andersonville noted that prisoners were "searched and robbed of everything valuable—watches, money, knives." He managed to save his watch by hiding it in his shoe. A novel written by a Southerner during the Civil War depicts a Confederate hero using his gold watch to bribe a Yankee guard. Other prisoners of war who escaped being robbed "bartered or sold their timepieces" while in prison.[30] Watches were signs and symbols of power, equal to pistols and rifles: in one portrait, an African American soldier poses with his pistol in his lap, a watch chain prominent across his chest.[31]

Soldiers' demand for watches is better illustrated by qualitative than by quantitative evidence. Quantitative data about the distribution of watches among soldiers during the Civil War are hard to come by. Social historian Mark Smith has looked at clock and watch distribution in rural and urban South Carolina, finding a rise in timepiece ownership between the 1840s and 1860s. Distribution of watches and clocks increased among members of the lowest wealth quintile, who would have been most likely to enlist.[32] There is scant quantitative data for watch acquisition in the North during the decade of the Civil War, or indeed any decade of the nineteenth century except the 1870s. During that time, about 9 percent of the factory-made American watches repaired by Union veteran D. E. Hoxie in Northampton, Massachusetts, had been manufactured and sold during the Civil War. The American Watch Company, later known as the Waltham Watch Company, made all fifty-one of these watches. Some of Hoxie's customers were Union veterans, whose watches may have been acquired during their war service. In one instance, Hoxie noted "37th" instead of the customer's name, which likely referred to the Thirty-Seventh Regiment, a regiment made up of soldiers from Hampshire County, Massachusetts (Northampton was the county seat). The unnamed customer owned a watch produced by the American Watch Company during the Civil War.[33]

Even without quantitative data, it is clear that military and civilian demand for watches bolstered the fortunes of retailers, wholesalers, and watch manufacturers. Many jewelers thrived during the war years. Consider the jeweler James E. Caldwell, who started a business in Philadelphia after the financial panic of 1837. Caldwell was an innovative retailer, introducing the one-price system in the 1840s, the same decade that dry goods merchant A. T. Stewart made his name for doing so. Just as the Civil War bolstered Stewart's fortunes (he supplied ready-made uniforms), so too did it improve Caldwell's, whose sale of watches skyrocketed. After the war, Caldwell reserved the entire first floor of his new "exceptionally large" jewelry store for watches; seven thousand square feet were devoted just to their display. A few years later Caldwell sponsored what was described in period literature as a "magnificent booth" at the 1876 Centennial.[34]

American watch manufacturers and importers prospered during the Civil War. By the close of the war, seven watch factories had been capitalized, while watch importers in every major city in the North were awash with business. Indeed, wartime demand for watches was the most important factor in the industry's lasting success. By the 1876 Centennial Exposition, American manufacturers quantitatively and qualitatively surpassed the English and the Swiss in the manufacture of watches. Few historians have addressed the short- or long-term effect of demand for watches and clocks during the war years. The only study of the economic impact of the war that mentions watch manufacture claims that "the war inspired a sizable watch market but one of short and indeterminate duration."[35] Evidence about the postwar market for watches demonstrates, however, that the sizable watch market inspired by the war was neither short nor of indeterminate duration. Not only was the Civil War the takeoff moment for the American watch industry, it also accelerated demand for pocket watches of all makes and origins.

The intense demand for watches during the Civil War provided the American Watch Company with the opportunity to break into the market dominated since the 1840s by the Swiss. A glance at the American Watch Company's fortunes over the course of the war years demonstrates the crucial role wartime demand played in its short-term fortunes and long-term success. In 1860, the company's treasurer reported that "very dull times" had almost "stopped . . . sales." The following year saw sales plummet more than 50 percent. The company was headed for trouble; production had slowed such that its employees were taking "half to a quarter of their usual wages." Prospects turned around, however, in 1862, and not just for the least-expensive versions of the American Watch Company's watches. According to the treasurer's report for that year, "Each month sales include[d] a larger proportion of the better watches than the last." Business was so good, with more than 20,000 watch movements selling, that it was decided to increase prices 25 percent, finding that "business had not fallen off on account of it."[36] Such wartime opportunism indicates how strong was the demand for pocket watches; even a substantial price increase did not dampen it.

As the war dragged on, productivity, output, and profits continued to mount. In 1863, the American Watch Company registered an enormous profit. It had manufactured more than 38,000 movements at the rate of 150 per day, none of which were exported. The treasurer reported:

> I have the pleasure to say the demand for our productions continues not only without abatement but in quantities we find ourselves wholly unable to supply. Indeed the trouble for the last year has been to divide the watches amongst our very numerous customers as to produce the least dissatisfaction. The several advances in price we have been partly compelled and under these circumstances partly

invited to make, have been paid with entire readiness, and although these prices have now reached a very high point the demand continues as brisk as ever.

His punch line: "If we don't supply the demand others will attempt it."[37] Indeed, others did. The Swiss exported 226,000 watches to the United States in 1865 alone, while the English exported nearly 30,000 watches a year over the course of the war.[38]

Even in 1864, after the American Watch Company ramped up production to 250 movements per day, five American competitors entered the business.[39] The new entrants included the National Watch Company in Elgin, Illinois; the Newark Watch Company and United States Watch Company in Newark, New Jersey; the Tremont Watch Company in Boston; and the New York Watch Company in Providence, Rhode Island. Supplying the capital for all five of the new companies were jewelers who were making fortunes by meeting wartime demand for American, Swiss, and French watches, as well as for buttons, pens, and other accoutrements. Chicago jewelry wholesalers financed Elgin's National Watch Company. Boston jeweler A. O. Bigelow financed the Tremont Watch Company. New York jewelers and watch importers supplied the considerable capital for the Newark Watch Company and the United States Watch Company. Giles, Wales and Company, a New York jewelry house that did well importing Swiss and English watches during the war, provided inventive watchmaker Don Mozart with capital to start Providence's New York Watch Company.[40]

None of these companies, except Bigelow's Tremont Watch Company, managed to get watches to market during the war. Developing the design of the watches, building the machinery, acquiring the materials, and hiring the skilled labor took the other companies three years, by which time the war was over.[41] This delay in manufacturing spelled the end of all but one of these enterprises because none of them could repay its investors quickly enough. The machinery and labor from these failed factories, however, formed the basis of new watchmaking enterprises. Among these companies' signatures were Springfield, Hampden, Rockford, Aurora, Lancaster, and Waterbury.[42] Wartime demand for watches provided capital for significant investment in what would become a thriving sector of the economy. Furthermore, because demand only momentarily dropped off after the war, these new factories were able to capture the American market, surpassing the Swiss and English in sales by the 1870s. The age of the American watch had dawned.

Of no small importance to the American watch's popularity were developments in watchcase making and decoration. High tariffs on watchcases dating to the earliest years of the Republic encouraged the growth of case-making shops, particularly in Philadelphia and other port cities. In the 1860s, most of these shops could only make ten cases per day, which hardly kept up with the output of watch factories or the efforts of importers. By simplifying the design of cases, mechanizing their production, and adopting new techniques based in Boss's spinning-up technique, it became possible for some factories to make as many as six hundred cases a day. Manufacturers also devised new methods to make gold-filled cases, as well as inexpensive cases from various metal alloys. Among the jewelers who pioneered this industry were Philadelphia's aforementioned James Boss, New York's John Fortenbach and Joseph Fahys, and Cincinnati's John Dueber. Although some artisanal case shops still operated in the middle of the 1880s, Fahys and Dueber consolidated the industry. Customers were offered a choice from a variety of patriotic, pastoral, technological, and sentimental scenes that could be engraved on the casing.[43] The manufacture of watchcases in standard sizes simplified the retail jeweler's work, while their decoration enhanced their appeal.

The healthy market for timepieces that developed during the Civil War grew to a vast size by 1900. After the war, the supply of watches continued to increase, reflected in the large inventories of watch manufacturers and importers, as well as innumerable catalogs and trade cards featuring watch movements and cases.[44] Dozens of American factories opened, closed, reorganized, all the while churning out watch movements and cases of various grades, makes, metals, and styles. In the early 1870s, the machine manufacture of cases for watch movements was introduced, further bringing the expense of watches down.[45] These machine-made watch movements and cases were widely distributed. For instance, although most of the American watches D. E. Hoxie worked on had signatures associated with the Waltham Watch Company of Massachusetts and the Elgin Watch Company of Illinois, he also repaired watches made by short-lived companies such as the Hampden Watch Company, the United States Watch Company, the Philadelphia Watch Company, the Tremont Watch Company, the Home Watch Company, the Metropolitan Watch Company, and the Union Watch Company.[46]

As the price of American watches dropped and their quality increased, imported European watches encountered competition. Nevertheless, a huge variety of imported watches still constituted more than half of the watches in use in the United States. Nearly two and a half million Swiss watches alone were imported into the United States over the course of the first decade after the Civil War.[47] Although the Swiss were reluctant to adopt the machine manufacture of watches, they were able to produce inexpensive watches well before other national industries owing to a number of factors, including tight control over labor costs and the introduction of an inexpensive, keyless watch known as the "proletarian watch."[48] The manufacture of English watches, long a staple in the American market, typically followed the artisanal system. The English, however, did very little to change the way they made watches, the kinds of watches they produced, or how their watches were distributed. As a consequence, their share of the market diminished considerably.[49] That Hoxie repaired watches with more than 120 different European signatures reveals the great variety of European-made watches circulating throughout the United States. What is more, 65 percent of his trade as a whole was in European watches, more than half of which were Swiss. The Swiss were not content with half the American market for imported watches. They responded to high tariffs and competition from American manufacturers by producing "Imitation American" watches. Hoxie repaired forty-seven of these watches, invariably noting their inauthenticity with the abbreviation "ImAm."[50]

The insistence of the Waltham and Elgin companies on the high quality of American-made watches was no sleight of hand. Waltham watches were granted the highest honors and prizes in trade competitions, most notably the era's world's fairs.[51] Of the first competition against Swiss watches, at the 1876 Centennial World's Fair, Waltham's treasurer reported: "Our triumph at Philadelphia was undeniably the most conspicuous of all American Successes." Even leading Swiss manufacturers admitted to the excellence of American watches.[52] Watches were among the first American-made consumer goods that actually were superior to those of European make. The robust American market for these watches stimulated the development of the industry. As Amy Glasmeier, a student of the international watch industry, has commented, "a complete reconceptualization of the watch as both a technology and as a consumer product was required." She explains that until the 1860s and 1870s, good watches had been expensive items, their cost correlated with the degree of their precision. But with innovations in production and design, well-running watches could be cheap.[53] Americans embraced these watches, not just as inexpensive novelties, but also as repositories of sentiments and tools for success.

Machine-made watch movements, like nearly all other manufactured commodities in the nineteenth century, went from factory or importer to wholesaler, who then sold them in lots to mail-order companies, large dry-goods merchants, and retail jewelers. The Boston jeweler William Bond and Son, for instance, received at least twenty different circulars listing prices and attributes for watches and clocks from exporters with offices in London, importers with New York City offices, and manufacturers sending dispatches from their American factories.[54] In large cities, some merchants specialized in watches, chronometers, and clocks, but it was typically jewelers who dominated the trade. In 1873, responding to the expansion of the retail trade in jewelry, especially watches, the Jewelers Mercantile Agency began operations, which included publishing a directory of wholesale dealers in "jewelry, watches, clocks, diamonds and precious stones."[55] A glance at one of its guides provides a sense of the extent of jewelry trade, since wholesalers might have served hundreds if not thousands of retailers in the United States and abroad. In 1884, a trade guide listed 90 clock manufacturers and wholesalers, 2 wholesalers of cuckoo clocks, 2 manufacturers of astronomical clocks, 3 of electric watch clocks, 2 manufacturers of tower clocks, and 248 watch manufacturers and wholesalers. Chicago was home to 27 watch wholesalers, Boston (the crucible of the American watch industry just thirty years earlier) to 18, and Philadelphia to 16. Eighty-three watch wholesalers clustered on New York City's Nassau Street and Maiden Lane, where a private police force paid by the Jewelry League was on patrol.[56]

As one historian of American pocket watches has pointed out, the jewelry trade's tendency to cater to the affluent made it a poor vehicle for the distribution of low-cost mass-produced goods, like inexpensive watches.[57] As a consequence, after 1880 most manufacturers who made inexpensive watches bypassed wholesalers in order to reach wide markets. The well-known Waterbury watch was given away with tobacco, suits of clothes, and newspaper subscriptions. It was sold so cheaply that eventually the company changed its name in order to shed the impression that its watches were practically worthless.[58] The Manhattan Watch Company sold its low-priced watches directly to retailers. The Keystone Watch Club Company started watch clubs in which consumers bought watches on installment; they attended weekly meetings where they made their payments and learned about the virtues of clock time.[59] The market for inexpensive watches was subject to satire; rural people especially were depicted as being susceptible to canny salesmen offering goods like wooden watches.

Americans did not have to visit a jeweler's shop to buy a watch. Getting one through the mail was easy: "Simply write that you want one, with your name in full and your post office address," invited the Independent Watch Company of Fredonia, New York. Many jewelers and retailers did mail-order sales, issuing catalogs and advertising in regional newspapers.[60] Richard Warren Sears's first foray into mail order was with watches: in 1886, he founded an eponymous mail-order watch company in Minneapolis, which he sold in 1889 for a small fortune. In 1893, Sears, Roebuck and Company debuted, selling by mail all sorts of consumer durables, including an extensive line of watches. Montgomery Ward, whose mail-order business started in 1872, sold many lines of American watches for men, women, and children, as well as watchmakers' tools and supplies and various types of clocks.[61] After 1900,

"How a Countryman 'Bought a Watch,'" cartoon by Holcomb and Davis, 1868, Picture Collection, New York Public Library.

the Illinois Watch Company of Springfield made the private label "Washington Watch" for Ward's and the "Plymouth Watch" for Sears. The largest retailers in the nation—Ward's and Sears—recognized the advertising value of *American* names when it came to watches.

As with other durable-goods manufacturers, productivity increased so considerably after the Civil War that watch manufacturers scrambled to maintain profits. After the mid-1870s, Waltham and Elgin annually manufactured at least a hundred thousand watch movements each. Although not as productive as Waltham and Elgin, other American watch manufacturers made tens of thousands of movements as well.[62] At first they tried to limit competition. The implementation of high tariffs reduced competition from imports. A watch trust limited production and fixed prices, but, as happens to all informal agreements to restrict trade, it collapsed.[63] Watch manufacturers also sought larger markets. Immediately after the Civil War, Waltham dispatched agents to Chicago, Canada, and the Pacific Coast in the effort to increase sales and build the company's reputation "as the leading watchmakers of the present time."[64]

Eventually, in order to dispose of their products at a profit, watch manufacturers had to rely on the core feature of mass marketing: differentiation of the product through branding and advertising. They branded their watches not only with company names, but also with unique signatures—the "Lady Elgin," Waltham's "P.S. Bartlett," or the United States Watch Company's "Young America"—that explicitly or inexplicitly indicated suitability for particular target markets. Elgin, according to a company history, sent out "goodwill men, known as missionaries, who visited retailers to talk about the product." By the 1880s, Elgin's advertising energies were surging: a bandmaster was hired to "build up a military band to give promotional concerts."[65] Elgin, Waltham, and other manufacturers also advertised their brands through a mixture of trade cards, circulars, almanacs, pictorial engravings, and newspaper notices. Rarely is the imagery urban; instead it tends to depict men near railroad tracks or men, women, and children in kitchens, parlors, and pastoral settings. In the 1890s, Ingersoll campaigned hard using national symbols to sell its watches. Flags and soldiers accompanied its "Yankee" watches, which its trade card promised was one of "the cheapest guaranteed watches in the world."

Michael O'Malley interprets Gilded Age watch advertisements as plumbing three central themes: they promise precision and accuracy, access to improved social status, and a merger of self with machine.[66] This last theme reflected not only aspirations to internalize clock time, but also anxieties about it that were consistent with ones expressed in the antebellum years. Fears abounded about mechanical timekeepers colonizing people's bodies and minds, as is evident from an 1899 theatrical poster showing a hired girl with clothespins in her mouth trapped in a stem-winder pocket watch that showed hours, minutes, and seconds. This hired girl, like Frank Baum's Tin Man from *The Wizard of Oz*, reflects the era's preoccupation with the perils of mechanization: would the human heart, itself a timekeeper, cease to beat?[67]

During the Gilded Age, pocket watches fascinated Americans, but as had been the case a century earlier, they were optional, rather than necessary,

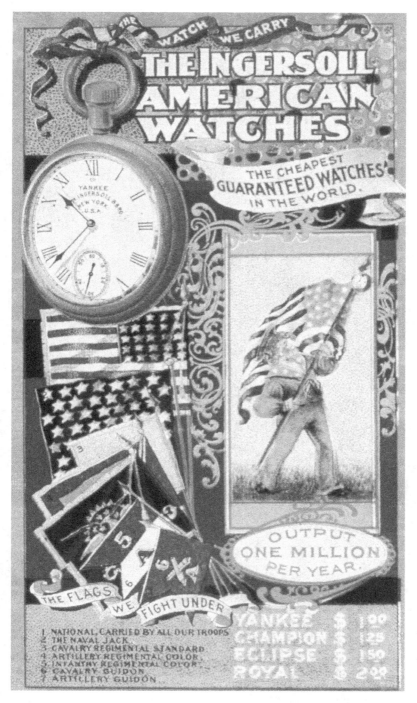

Trade card for Ingersoll American Watches, ca. 1898–1900, Warshaw Collection, National Museum of American History, Smithsonian Institution. No doubt "Yankee watches" accompanied the flags and soldiers that went to Cuba and the Philippines during the Spanish-American War.

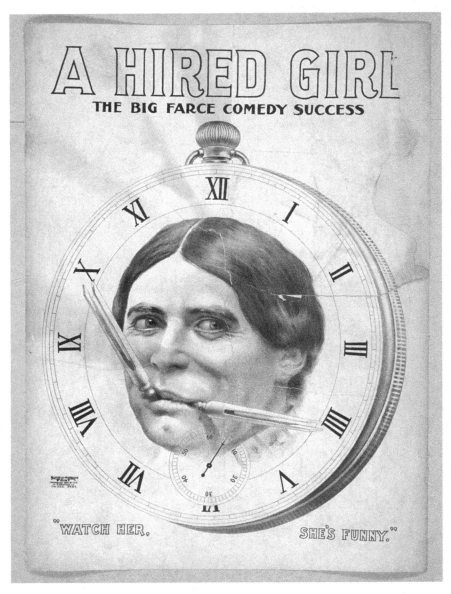

Theatrical poster for *A Hired Girl*, 1899, from the US Printing Company, Cincinnati, Theatrical Poster Collection, Library of Congress. The glint in this hired woman's eyes promises the staging of a play sure to disrupt class, gender, and time hierarchies. Contrast this image of a working-class woman as a captive of mechanical time with that of an upper-class woman who appears to effortlessly inhabit the realms of natural and mechanical time in the Waltham Watch advertising poster reproduced on page 86.

accoutrements.[68] In fact, they caused a lot of anxiety. Even so, many pocket watches ran very well and were increasingly easy to set and consult: they were useful to men and women intent on internalizing clock time. The repairs to watches, as well as their casings and materials, offer some clues to these time-keepers' use and meaning. Owning watches made Americans nervous, and not just because watches constituted a "form of mobile capital."[69] Why was so much care taken to pin them to dress bodices, secure them with chains, keep them in pockets, case them, stow them in specially constructed stands, if not because a number of anxieties coursed through ownership?

Almost as soon as the Civil War ended, publishers brought out a variety of watch repair manuals, some for uninitiated but eager watch repairmen and others for people "Not Finding it Convenient to Patronize a Horologist." As H. F. Piaget, a charismatic watchmaker, wrote, demand for his guide *The Watch* increased "after the war ceased."[70] Period sources recommended maintenance from a skilled repairman at least every other year. For instance, an 1879 mail-order catalog for watches admonished readers in a section titled "How to Use a Watch" that "no watch should go longer than eighteen months without cleaning (small ones in particular), as the oil will by that time be exhausted and the pivots injured; once a year would be better." Watch repair shops, like D. E. Hoxie's in Northampton, and repair departments in large jewelry enterprises could be found in most American towns and cities well into the twentieth century.[71]

Jewelers and watch repairmen were not necessarily trusted allies. Mark Twain's short sketch "My Watch: An Instructive Little Tale" satirizes his attempts to repair his watch, describing to a knowing audience his "heavy hearted" visits to a handful of watch repairmen, whose competence was doubtful. Hyperbole notwithstanding, Twain gave voice to the anguish many Americans experienced when their pocket watches failed to run. The narrator opens with a description of his "infallible" watch, which he "set by guess" after it ran down one night. The next day, he "stepped into the chief jeweler's to set it by the exact time," when "the head of the establishment took it" and "proceeded to set it." But the jeweler found that the watch was slow and fixed it so that it ran faster and faster, leaving "all the timepieces of the town far in the rear." Twain enjoys imagining his watch running fast: "It hurried up the house rent, bills payable, and such things." So, he explains, he took it to another watchmaker, who "eagerly pried it open" with "a look of vicious happiness." This second repairman "cleaned and oiled and regulated" the watch such that it "ticked like a tolling bell." Here Twain contrasts the measure of time shown by the watch with measures of occasion sounded by bells, as well as the visibility of time on a dial with the audibility of time when sounded on a bell. With the watch reverting to this older temporal system, its owner "began to be left by trains," "failed all appointments," and even missed dinner. He "drifted back into yesterday, then the day before, then into last week" until he had a "sneaking fellow-feeling for the mummy in the museum."[72]

The attentions of other watchmakers give Twain the opportunity to invoke still other temporal markers, while lambasting the amateurism of watch repair services. The narrator explains why a visit to a third and then a fourth watchmaker was necessary; the watch ran sporadically, and each time it started, "it kicked back like a musket." A fifth watchmaker failed to give it "a fresh start" but instead contrived to have the hands "shut together like a pair of scissors." After suffering the sixth repairman, "everything inside [the watch] would let go all of a sudden and begin to buzz like a bee." The hands spun "round and round so fast that their individuality was lost," so much so that the watch face looked like a "spider's web." Here Twain juxtaposes the natural time of bees and spiders with the artifice of mechanical time. The tale closes with the owner in despair as he weighs the expense of the watch—two hundred dollars—against the seeming thousands already spent on its repairs. When he recognizes yet another watchmaker for "an old acquaintance—a steamboat engineer of other days, and not a good engineer either," the narrator reaches the end of his tether. He kills and buries the erstwhile engineer, who had confidently explained that the watch made "too much steam." And then Twain delivers the moral of the story, one with which many Americans likely agreed: "A good watch was a good watch until the repairers got a chance at it."

The litany of possible malfunctions would have resonated with Twain's readers, who had plenty of reasons to visit watch repairmen. The bent hands, shattered glass, lost jewels, broken mainsprings, and myriad other damages that watchmakers ameliorated often ensued from mistreatment. Broken watches have a long pedigree; eighteenth-century watch owners were frequently disappointed by the unreliability of their timepieces.[73] Watch owner's manuals suggested that watches be kept near the same temperature at all times, that they remain in "as uniform position as possible," and that at the same hour each day they be wound with a key that exactly fit and was "perfectly clean."[74] Pivots, levers, escapements, wheels, wheel trains, hairsprings, fusees, clicks, ratchets, and pinions frequently broke. Repair accounts reveal little harmony between owner and machine. Even with all the improvements in watch manufacture and the ready access to competent repairmen, watch owners were not especially well attuned to their timekeepers. Upkeep seems to have been a mysterious rite rather than a daily duty, and carelessness is writ large through watch repair accounts. Rather than domination by the machine, it is dominance over the machine, a destructive sort that resulted in broken hands and stems, that was the rule.

Comic "Jeweler" valentine, ca. 1865–80, courtesy of the Library Company of Philadelphia. The ditty beneath the pensive figure in the valentine gave voice to the suspicion that the work of such repairmen was inept: "A crowbar, or pickaxe, a more fitting tool / Would be in your hands, you ridiculous fool."

The frequency of repair work to the mainspring, which in uncoiling at a steady rate provided the watch with its going mechanism, leads to the conclusion that incorporating a watch into daily life proved difficult. The mainspring, which was a ribbon of steel and therefore unlikely to break on its own, loses tension with sudden changes in temperature or with overwinding.[75] While a handful of American owners may have carried their watches with them as they moved from warm interiors to cold exteriors, the fact that the watch was usually close to the body, in a pocket, under a coat, would have protected it from such extremes. It is more likely, then, that owners overwound their watches, in part owing to lack of familiarity with the mechanism, in part to anxiety lest it stop ticking. Watch guides endlessly repeated maxims concerning the need to wind one's watch every day at the same time, that is, with clockwork efficiency. And yet broken mainsprings suggest more frequent winding was the rule. Even after 1900, neither mainsprings nor watch owners were reliable: a 1901 advertisement for the "Veritas Model" of the Elgin "Long Run Watch" promised, "It has the longest, widest, strongest mainspring ever used in a watch." Moreover, "[It] makes amend for the poorest memory. There is no probability of winding down."[76] In conclusion, what one newspaper reporter described in 1880 as the "mental agony" and "anxiety" caused by broken watches was widely known. It was said that "every man, in his own breast, doubts his watch."[77]

The materials out of which watches were made also provide a window into their meanings and uses. Over the course of the nineteenth century, the silver watch connoted frugality, punctuality, thrift, sobriety: silver watches were not counted among luxury items. Nineteenth-century fiction is replete with physicians, kind old men, surveyors, and customs house officers carrying trustworthy silver watches: "a deliberate man," Dr. Gaines "drew out his silver watch to look at the time"; a druggist took "out his silver watch with a professional air"; "an old land-surveyor" who carried a compass, read two newspapers daily, and seldom spoke "wore a double cased silver watch, that kept time to a second with the sun all year round."[78] Gold watches, however, conveyed mixed messages. In fiction, they signified an inability to marshal resources and spend wisely, excessive attention to outward appearance, and pretensions both dangerous and aggressive. One tale's villain, Major Morrow, is described as a "satanic" man with a "great brawny frame" and "keen and cruel grey eyes," who was "fashionably dressed" and "seemed not a little vain of a magnificent gold watch."[79]

A gold watch could assume benign qualities when acquired as an heirloom, gift, commemorative object, or sentimental memento. In the 1880s, the custom of giving gold watches to commemorate long service and retirement began; ironically they were bequeathed at the moment when they were no longer needed.[80] Women usually received watches, many of which were gold, in honor of entry into society, an engagement, an anniversary, or the birth of a child. The watches often "bore special messages of commemoration engraved on the case," according to Carlene Stephens, "or featured dial and movement finishes customized for the recipient."[81] Records from Hoxie's Northampton repair shop indicate that women owned a disproportionate number of gold watches: four out of five watches they brought in for repairs were gold.[82] Here the gold watch—with its many ambivalent associations and meanings—conveyed a competing tendency, viewing women's time as a symbolic, rather than a real, resource.

The popularity of the hunter case suggests that Americans were not tightly bound to their watches. Hunters were closed cases and temporarily protected the watch's movement from sudden changes in temperature and precipitation, from blowing debris, and from rough treatment. With the watch face covered, attending to the time was a deliberate, even ritualistic, act: a man would reach into his pocket, pull out the watch, open the case, read the time. Since he could not leave it dangling by the chain once he ascertained the time, he then had to close the case and return the watch to its pocket. By the same token, a woman would unpin her watch from her dress or remove it from a small pocket in her waistband, then lift the chain securing it around her neck, before she could open the case to discover the time. And again, she would have to repeat these steps in reverse. Women's watches were sometimes parts of decorative brooches. Gradually open-face cases became more popular than the hunters.[83] That Americans' need to know the time was simply not as pressing as it would become in the twentieth century is the consequence of both external and internal factors: timetables and schedules were estimations, rather than absolutes; mechanical time (clock time) had not yet penetrated people's consciousness to an extent that made it necessary to check the time instantly or repeatedly; and indeed, abundant signals like bells and whistles forestalled the need for a thorough internalization of clock time.

The materials and casings of watches mediated between accumulating imperatives to live by the clock and the strongly held orientation to other temporal measures rooted in nature, national history, religion, and the life cycle of birth, coming-of-age, marriage, retirement, death. They softened the insistent, challenging aspects of one of the most modern of all objects—the personal timekeeper. The watch bonded generations; it signified continuity and unbroken ties between fathers and sons, mothers and daughters, husbands and wives. Watches reinforced multiple temporal senses: in addition to mechanical time rising out of the movement's work, the casings communicated senses of the past, the present, and the future; the materials signified wealth and aspirations. The admixture of pocket watches' value as investments (precious metals, jewels, fine machinery), as objects conferring status, and as conveyances of sentiment and memory augmented their timekeeping utility, and perhaps compensated for their unreliability.

Advertisement for the Waltham Watch Company, ca. 1900, Advertising Ephemera Collection, Baker Library Historical Collections, Harvard Business School. In this color poster from the turn of the century, the time and tide are equated with the form of a watch and the body of a woman.

Pressures to become time disciplined could be felt everywhere in the 1870s and 1880s, but their force was not so overwhelming as to compel large segments of the community to comply through the acquisition of watches. There were other avenues and technologies that enabled modern time discipline, such as public timepieces and domestic clocks, which were cheaper to purchase and required less maintenance than watches. The desire for the correct time is but one important manifestation of the rising tide of clock-time orientation over the course of the nineteenth century. Pocket watches were devices that could grant individuals control over the passage of time, the ability to coordinate events within time, and access to the experience of mechanical time itself. They heightened the potential for efficiency, synchrony, and simultaneity.

Knowing the extent of the distribution of watches during the public clock era would illuminate the role they played in fostering modern time discipline. Hoxie's repair records provide a sense of the distribution of watches in a small northeastern urban community toward the end of the nineteenth century.[84] As a proportion of the population, it appears that no more Northampton residents owned watches in the late 1870s and early 1880s than did those of Long Island or Connecticut in 1800. Certainly the sheer number of watch owners increased across the nation, and it might be that a larger proportion of the residents of large cities owned watches than those of smaller cities, towns, and villages. About the watch owners in Northampton, however, a few traits stand out. First, there was not a direct correlation between wealth (or lack thereof) and watch ownership. Second, although owners were disproportionately men, many more women in 1880 than in 1800 owned watches.[85]

It is all but impossible to estimate using any but production and import data how many Americans owned watches after the 1870s. Probate inventories, the source used for colonial and antebellum estimates, do not provide an accurate snapshot of real property ownership after 1860. Sales figures for durable goods leave out the secondhand market and heirlooms, let alone the timepieces that people already owned. Evidence about domestic production and importation of watches suggests that during the last two decades of the nineteenth century there was a significant increase in watch ownership in the United States. American and foreign watch manufacturers improved their productivity, and American demand for watches increased.

The 1880s were a turning point in the marketing of watches, as the fortunes of the Waltham Watch Company demonstrate. Its treasurer commented in 1882: "It is safe to say that we could have doubled our sales during the last two or three years if we could have doubled our product." In 1883, when the economy slowed down during one of the Gilded Age's many busts, he explained, "All the cheaper goods sell readily. It is only the higher grades which hang heavily on our hands. Unfortunately it is just those on which we make most profit." The economic depression may have deepened in 1884, but the treasurer was pleased to report that Waltham "never made so many good watches nor made them so cheaply as now." The company actively sought foreign markets with displays at the New Orleans Cotton States Exposition (1884) as well as an International Inventors Exhibition in London in 1885. By 1887, Waltham was reporting that it had sold more watches than ever before in its history. Profits were large enough that the company bought a street railway in 1888 and had its factory electrified in 1889.[86]

By the early 1890s, the nation was saturated with watches. The volume of watch manufacture and the import-export trade was so large, prices for movements and casings so low, innovations in design so numerous, and quality so superior that it is likely that at least half the adult male population owned watches. Waltham and Elgin together made at least one and a half million watches in 1892 before a major economic setback cut production in half. Other American watch companies, like Hampden Watch Company, also had high levels of output.[87] Millions of Swiss movements were imported to the United States. Heated competition among manufacturers pushed the prices of watches downward. By 1892, a dollar could pay for an American-made watch ("the watch that made the dollar famous," as the Ingersoll Watch slogan claimed).[88] Not only were watches widely available, they were, for the most part, well made. After the 1890s, rare is the cultural reference to watches as mystifying and troubling.

The absorption of more and more new watches by the American market in the 1880s and 1890s was as much the consequence of accumulating pressures to follow clock time and the increasing ease of doing so, as of lower prices for watches. The correlation between rising supply and demand might lead to overlooking other sources stimulating demand for watches besides falling prices. Falling prices for consumer goods can stimulate demand, but the correlation between the two is not perfect; that is, falling prices do not entirely account for rising demand. Widespread synchronization between pocket watches and public clocks did not happen until the 1890s. By this decade, agreed-upon time standards were widely disseminated, many clocks were synchronized with reliable timekeepers, and the technology through which time signals could be sent had matured. With these developments, pocket watches would become indispensable features of modernity. A network of synchronized public and personal timekeepers adorned with eagles and flags would allow an increasingly time-obsessed population to strive for modern forms of time discipline. Pocket watches enhanced the importance of standard time and its conveyance, public clocks.

CHAPTER 4

Noon, November 18, 1883

The Abolition of Local Time,
the Debut of a National Standard

This improvement in watch hands is designed to enable the wearer to see at a glance the different times of the place he is leaving and the place of destination, or to enable him with one watch to keep both standard and local time. . . . This improvement will be appreciated by all travelers, and by others who are obliged to differentiate time. It is possible that this small device may go a long way toward introducing standard time.

— "Bell's Improvement in Watch Hands," *Scientific American* (1882)

n November 1882, a tinkerer and wealthy investor, John W. Bell, received US Patent 267,824 for a pair of coupled watch hands, "designed to enable the wearer," as the patent proclaimed, "to keep both standard and local time." Although American railroads would not introduce standard time until a year later, the problem of showing multiple times on one timepiece was one that inventors had grappled with for several decades. In 1865, a patent was issued to New York watchmaker A. W. Hall for an "improvement in universal time-pieces," which with several hands would show the local time and the mean time. Somehow with the aid of a paper dial inserted over the watch face, Hall's "improved" watch dial would also indicate the time in major American cities. A decade or so later, the Canadian Sandford Fleming, who in the 1880s would campaign for a uniform system of time-reckoning throughout the world, also filed applications for US patents on watches that would address the lack of congruity between local and railroad times.[1] Bell's was a far more elegant solution to the problem of visualizing multiple times than Hall's or Fleming's. Its advantage was that by 1880 establishing one national standard of time had become a felt need in the United States, whereas in the 1860s and 1870s it was not.[2]

Not a month after Bell's patent was issued, the journal *Scientific American* praised "Bell's improvement in watch hands," with an article, a diagram, and the speculation "that it is possible that this simple device may go a long way toward introducing a standard time."[3] This was mere puffery; Bell's coupled pair of watch hands was neither novel nor catalytic. The process of standardizing time, which took nearly fifty years, depended instead on strenuous technological and ideological efforts to fashion an understanding of mechanical time as a social convention rather than as a reflection of a natural fact. It was also grounded in the rising awareness of simultaneity, that is, of the unfolding of events in distant places at the same time. Although the railroads had the audacity to declare national standard time a fait accompli one Sunday in November 1883, the nation's public clocks, time distribution systems, pocket watches, and sympathy for clock time not only helped make it happen, they assured its longevity.

Before coordinated clocks and a national standard of time, the experience of simultaneity was rarely apprehended. Even the telegraphic transmission of railroad baron Leland Stanford swinging at, but missing, the golden spike that would complete the transcontinental railroad in 1869 did not foster a sense of simultaneity. Instead, what Michael O'Malley has called "the Babel of local times" generated controversy about the actual time of the event. Indeed, with the numbers of clocks in public spaces increasing, it was noted that year, "they are sometimes the source of very great annoyance, particularly when there are several in one and the same place, from the fact that they rarely correspond with each other."[4] But even if clocks remained out of kilter owing to the imprecise operations of their works, after the introduction of national standard time the possibility of apprehending simultaneity beckoned.

By disseminating standard time, bells, time balls, and clocks brought a widely dispersed and heterogeneous population into one temporal frame. Standard time and coordinated clocks instantiated in another way, in addition to the calendar of days, weeks, months, and years, the apprehension of simultaneous experience. Different types of timekeepers emphasized different sorts of simultaneity. The time ball drew attention to the "now"—a

J. W. BELL.
WATCH HANDS.

No. 267,824.   Patented Nov. 21, 1882.

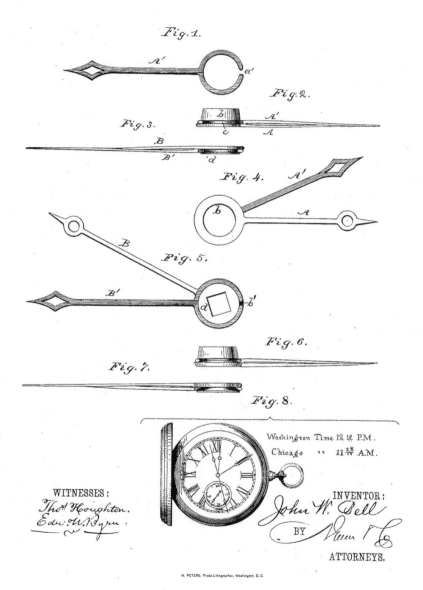

WITNESSES:
Thos. Houghton.
Edw. W. Byrn.

INVENTOR:
John W. Bell
BY
ATTORNEYS.

N. PETERS. Photo-Lithographer, Washington, D. C.

J. W. Bell's US Patent application for watch hands, 1882.

shared moment in which all onlookers were bound to each other. Bells served a similar function. Although their aural cues did not require a suspension of attention and activity, let alone the gathering of a crowd, as with the drop of time balls, both behaviors frequently happened when bells rang. Clocks and watches, silent keepers of the time, always available for consultation, spoke to a slightly different set of opportunities: rather than wait for the ball to drop or the hammer to strike, Americans could look upon their timekeepers at whatever moment they felt so compelled, to ascertain for themselves the minute and hour. A sense of simultaneity came when they could map themselves onto the same grid of clock time in which a distant event—an assassination, a first pitch, an explosion—occurred. The system of national standard time meant to coordinate clocks allowed for the cultural construction of simultaneity, a state of awareness necessary for both the fluorescence of nationalism and modern time discipline.[5]

The absence of coordinated clocks deeply dismayed the first person to provide an account of the history of standard time, its chief architect, William F. Allen, and not because he cared a whit about simultaneity or nationalism, though he did care a great deal for clock-based discipline. The description of Allen as the "time shuffler of all time shufflers" fits perfectly.[6] In 1873, Allen began editing the *Official Railway Guide*, a monthly collation of railroad timetables and maps prefaced with railway industry news. He also served as secretary of the North American railroads' General Time Convention, which facilitated the coordination of railway schedules. According to Allen, not only did his efforts introduce a practical system of standard time, but they also fomented "the abolition of local time."[7] A deeply ambitious man, Allen was determined to secure his place in history as the deliverer of standard time, the adoption of which he described as "an event which is likely to be noted in the history of the World for all time to come." In his drive for recognition, he obscured the role of organizations other than the railroads and actors other than himself in the standardization of time.[8] Historians have since brought attention to the part played by scientific organizations, especially those whose practice required time-stamped observations, such as meteorology and astronomy, in bringing standard time into widespread use.[9] While tracing the process through which mechanical time was standardized in the United States, I consider the imaginative technological fixes ranging from watch hands to time balls that emerged as multiple times became ever more unwieldy.[10] In doing so, I attend to the shift away from local standards of time to a national standard time. With local time gone, the way was clear for the public clock era to flourish, and with it, for modern time discipline to deepen its influence.

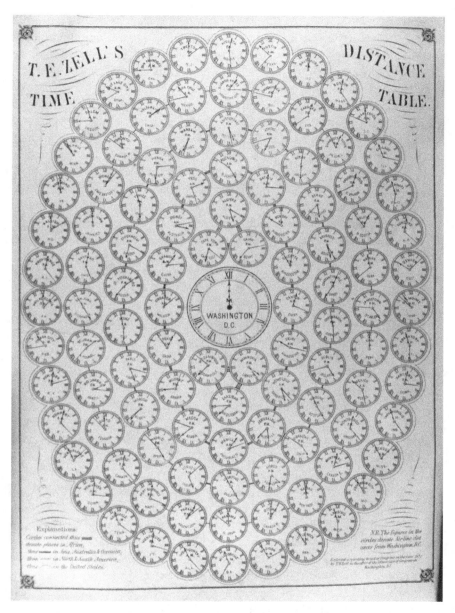

"T. E. Zell's Distance and Time Table," 1871, Warshaw Collection, National Museum of American History, Smithsonian Institution. Zell's elaborate time and distance table denotes the innumerable local times in effect through the 1870s. It also places at the center Washington, DC, time.

After the Civil War, railroad times were something to be reckoned with, but they still did not yet set the standard of time for American communities. It is estimated that ten US cities set the time for more than half the hundreds of railroads running in 1874. In 1883, just eight American cities set the time for two-thirds of the nation's railroads; the remaining third followed forty-one different civic time standards.[11] In many cities, railroad and local times were different; thus, clarification about what time a clock displayed was often necessary. During the 1870s, orders were placed with the leading clock manufacturers for timepieces bearing their engraved names and standard of time, for example, "Rail Road Time," "City & Train Time," "City Time," and "City Timekeeper."[12] In some places, jewelers were city-appointed time-keepers, but most often they simply claimed the authority by referring to their fine instruments and scientific expertise as justification. One Los Angeles jeweler's trade card asserted that because he took "transit time . . . from the sun weekly," his time was the city standard.[13]

Before the national standard was introduced, some clocks were engraved with the imprimatur "standard time."[14] Jewelers were the most likely to claim their time as the standard. Throughout the 1870s, the words "standard time," "true time," "meridian time," or even the uninflected "correct time" were engraved or painted on clock dials and glass as far west as Montana.[15] Jewelers went to great lengths to provide the correct time. For instance, in 1869, St. Louis jewelers donated a "valuable clock" to Washington University, which was hung inside the entrance to its sole building for the benefit of community timekeeping: "The public generally can step within our door and obtain the correct time." In exchange for the gift of the E. Howard clock, Professor C. M. Woodward agreed to "regulate it by astronomical observations" to mean solar time. The professor wrote to the city's jewelers with gratitude for their "valuable contribution to the cause of accuracy." He admonished them to "rigidly adhere to this standard and so secure to this city, what it has so long and so much needed, *uniform* and *correct* time."[16] Seeking accuracy, jewelers in other cities subscribed to observatory time services. The expense of such services was such that jewelers like Charles Teske in Hartford, Connecticut, tried to persuade their local city councils to foot the bill.[17]

In addition to conforming with scientific measures of the time as a means to bolster their claim to timekeeping authority, some jewelers ordered clocks that showed both local and railroad times. Clocks such as these were equipped with an extra minute hand, typically painted red, that indicated railroad time and a black minute hand that showed local time. Near the end of 1871, for instance, the jeweler Edward Vail of La Port, Indiana, ordered an E. Howard No. 2 regulator with an extra minute hand painted red. Nearly a decade later, the French Atlantic Cable Company of North Eastham, Massachusetts, placed a request for an ordinary clock "with two sets of hands to keep French & American time."[18] These dual-handed clocks accustomed viewers to multiple standards of time, but they also underscored the fact that railroad time, while important, was not preeminent. Railroads also ordered clocks with two sets of minute hands, thus acknowledging the existence of another standard of time than their own. For instance, the Boston office of the New York and New England Railroad ordered thirty-one clocks with an "extra minute hand painted Red for New York" for its stations.[19] None of these clocks are extant, likely because the extra minute hands were removed in the aftermath of the introduction of standard time.

To further appreciate the somewhat tenuous status of railroad time, consider timekeeping in Chicago, the nation's preeminent railroad city. In 1867, the Chicago Common Council designated Giles Brothers, a Chicago firm involved in the wholesale and retail watch, clock, and jewelry trade, as the provider of city time. A few years later it changed course, granting a contract in 1870 to the Chicago Astronomical Society, owners of the Dearborn Observatory, to supply the city with the time. To distribute the time, the Common Council approved illuminated dials for the cupola on the city's courthouse. In the meantime, the privately owned Fire Alarm Telegraph Company connected a pendulum clock to the courthouse bell so that the hours would be rung regularly, a practice that began in January 1871. That summer an E. Howard tower clock was installed in the courthouse. With the basis of a system of standard city time in place, the observatory planned to install "a system of synchronized clocks for Chicago railroad depots and business establishments."[20]

The Great Fire of 1871 laid these plans to waste; the city did not have the funds to pay for a contract with the observatory (which went unscathed in the fire), nor was there money for clock works and dials for its public spaces. Local jewelers stepped in to provide the city with authoritative time; to do so, a few subscribed to the observatory's time service. Just two weeks after the fire, the jeweler J. B. Mayo sent detailed instructions to E. Howard's Boston office for two clocks proclaiming "J. B. Mayo & Co. Time, Chicago." The firm's business cards explained, "Our chronometers are the standard time of the city of Chicago."[21] Giving voice to an aspiration more than a fact, this motto nevertheless underscores the competition for the privilege of setting the standard time in a city that was defined, above all else, by railroads.

In cities and towns, newspaper clocks competed with those of railroads and ambitious jewelers for civic attention. Well-demarcated senses of past, present, and future as defined by calendars and clocks define what is "news"; in working within these temporal frames, news organizations help to constitute national identities as well as narrative possibilities.[22] Publishers were

Trade card for J. B. Mayo, Chicago, ca. 1871–72, ICHI-65124, Chicago History Museum.

"anxious to impress the public with the civic spirit of their offerings," so they chose prominent locations for their buildings. In New York City, the *New York Times*, the *New York Sun*, the *New Yorker Staats-Zeitung*, and the *New York Tribune* situated themselves near city hall, whose four-faced clock sat atop a tower beneath a cupola. Here these newspapers competed with the city for authority over the time. The pioneering skyscraper designed by Richard Hunt Morris for the *New York Tribune*, then the tallest occupied structure in the city, was crowned with an illuminated four-faced clock in 1875. An electric clock system attached to its tower clockworks synchronized the building's clocks inside and out. That year, the *Evening Post* installed in its offices a marble dial clock in a black walnut case. Two years later, in 1877, the *New York Times* erected a double-dial clock on the façade above its building's entry: on both dials gilt letters proclaimed "N.Y. Times." The *New York Clipper* in 1879 raised a post clock in front of its offices: the forty-inch iron dials painted black announced "Clipper Time."[23]

In 1876, as the clocks of jewelers, railroads, and newspapers competed for authority with one another in various American cities, a new clock and bell for Independence Hall claimed national if not transcendent authority for the time. The Seth Thomas striking tower clock made explicit the inviolable link between mechanical and historic time. It was nearly a foregone conclusion that the nation's centennial would be celebrated with a world's fair in Philadelphia. As the year approached, new commercial buildings and clocks overshadowed the Old State House and its timepiece. A large clock topped the entryway to Fox's New American Theater, erected in 1870 at Chestnut and Tenth Streets. The city's Lincoln Market (1871) at Broad and Fairmont Streets boasted a clock tower. The Pennsylvania Railroad Company installed ten large clocks in its offices and depots during 1871 and 1872. In 1874, the Guarantee Trust and Safe Deposit Company moved into a new building with a tower clock. The next year, John Wanamaker acquired the old Pennsylvania Railroad depot and freight sheds, complete with clock tower, at Thirteenth and Market. After holding a religious revival in the depot, he opened an eponymous department store in 1876, for which he acquired a new Seth Thomas tower clock in 1880. These and Philadelphia's other commercial clocks appropriated national symbolism: a large eagle is draped over the clock on Fox's theater; flags fly near the clocks on the Lincoln Market and Guarantee Trust building; Wanamaker's clock was unveiled on the Fourth of July.[24] In each case, commercial entities borrowed the national authority clock, eagle, and flag had accrued over the preceding decades.

The Old State House reclaimed its pivotal position as authority for clock time in Philadelphia in 1876, the same year that the building's status solidified as a symbol of national origins. That year, Henry Seybert, a native son and reclusive heir to a large fortune, gave the city of Philadelphia a magnificent Seth Thomas tower clock, a 13,000-pound bell, and four nine-foot dials with gilded hands and figures. Having spent the 1820s and 1830s in Paris, Seybert had not witnessed the installation of the state house's tower clock in 1829, nor did he hear the Liberty Bell ring on state occasions before it cracked in 1846. In France during those decades, as Alain Corbin recounts, "disputes caused by national peals were legion," and surely Seybert often heard the ringing of bells even if he remained unaware of the controversy they sometimes caused. His motives for the gift to Philadelphia were largely personal. Seybert recounted that at a séance, his deceased mother commanded him to give a clock and bell for Independence Hall. Following her spirit's wishes, the terms of his gift called for the inscription of his family's personal names on the clock and bell. The practice of inscribing bells as a way to honor cherished members of the community was popular in France and not unheard of in America either. Members of the Common Council, attentive to the communal and national importance of the Old State House timekeepers, rather than to Seybert's personal affairs, refused the terms of the gift. Seybert eventually relented, agreeing to an inscription on the bell that read: "Presented to the city of Philadelphia July 4, 1876, for the belfry of Independence Hall, by a citizen."[25]

Fox's New American Theater, Philadelphia, 1870, Print and Picture Collection, Free Library of Philadelphia. The large clock above the entryway to Philadelphia's newest theater in 1870 was just one of many public clocks to go up after the Civil War.

The new clock and bell debuted in 1876 at the stroke of midnight on July Fourth: the bell, made of cannon from both sides of the Revolutionary and Civil Wars, struck thirteen peals. Then, "all the bells and steam whistles in the city joined in the sounds of rejoicing and fireworks and firearms made the noise tenfold louder." The new Independence Hall clock and bell assumed a central place in the daily lives of many Philadelphia residents. As one newspaper opined: "It would be impossible for us to do without it. The four faces of the State House timepiece are as familiar as the features of intimate friends; and the same may be said of its voice." Independence Hall's 1876 bell also marked important moments of national time. In 1888, an ordinance was passed requiring that the bell toll on New Year's Eve and July Fourth, in each case to "commemorate another year of American Independence."[26]

The exhibition of the Liberty Bell, which had been mute for three decades,

Centennial Fourth—Illumination of Independence Hall, Philadelphia, 1876, wood engraving, LC-USZ62-50709, Prints and Photographs Division, Library of Congress. This engraving showing the illumination of Independence Hall's new clocks and a parade streaming behind the Liberty Bell illustrated a widely distributed account of the nation's centennial celebration in *Harper's Weekly.*

at the Centennial Exposition the same year that Independence Hall received its heavy bell and fancy new clock reconfigured the meaning of historic time. The new hour's bell was itself a much heavier replica of the Liberty Bell.[27] With each decade of the nineteenth century, American bells had grown more numerous, larger, and heavier. Big bells evoked Europe's church towers, as well as the monasteries that preceded them. Their triumphal installations, however, commemorated independence from Europe, liberty from the past, and democracy's triumphs. What is more, as creations of advanced steel and metal casting technologies often rung by automated systems, these bells were manifestations of innovations in industry and science. By 1900, at least 750 American cities relied on a time circuit to ring their bells on a daily basis, usually announcing noon and nine at night.[28] Bells were central to the diffusion of information necessary for modern time discipline because they facilitated the distribution of clock time. Despite its silence, the preeminent bell in the nation, the Liberty Bell, bundled meanings associated with the nation's history as well as with the increasingly synchronized routines that characterized the day-to-day life of its citizens.

A great variety of the world's fair exhibits in 1876 showcased American-made watches, clocks, bells, and clock systems, as well as those manufactured by European competitors. Seth Thomas's "great clock," with its fourteen-foot-long pendulum, hung above the east door of Machinery Hall; an electric circuit connected it to clocks throughout several buildings' enormous exhibition spaces. The striking clock was attached to thirteen bells together weighing twenty-two thousand pounds: each of the bells represented one of the thirteen original states. Watch manufacturers Elgin and Waltham each exhibited their lines at the fair, enjoying their triumph over Swiss and English competitors in various precision trials meant to test the watches.[29] Each company's exhibits were more than platforms for these competitions; they sought to inspire fairgoers by demonstrating the possibilities of synchronicity. Despite the successes of American watch manufacturers, the impressive design and function of Seth Thomas's "great clock," and Independence Hall's state-of-the-art striking tower clock, the many dimensions of modern timekeeping had yet to coalesce. The most urgent need was for a reliable system to disseminate the time, but even more so, for an agreed-upon time standard.

What with the proliferation of various signs and symbols of clock time at the Centennial Exposition, along with the attempts of jewelers, railroads, and newspapers to set local time standards, it is no surprise that the era's leading communications firm, the Western Union Telegraph Company, and the United States government saw fit to enter into the "true time" business. In the 1870s, with the exception of one Light House Department order, the only federal government orders for E. Howard clocks were meant for Washington, DC: several interior clocks were purchased for the Government Hospital for the Insane and the Treasury Department, while an astronomical clock for the US Naval Observatory was requested in 1873.[30] Two years later, Western Union arranged to sell the US Naval Observatory's time signals, which it received in exchange for access to the telegraph wires. To announce

Western Union Telegraph Company headquarters, New York, ca. 1888, New-York Historical Society. Visible are the clock and time ball on the headquarters of the premier telegraph company in the United States.

this service, in 1877 Western Union erected a time ball on top of its New York City headquarters and announced that daily, except Sundays, the ball would drop precisely at noon. A seeming wonder of modern technology, the Western Union time ball was telegraphically connected to the clocks, transit telescopes, and scientific expertise housed in the nation's premiere observatory, that of the US Navy.[31] Western Union's time service provided timekeepers, be they jewelers, scientists, railroad managers, or astronomers, with US Naval Observatory time, which could then be translated into local time by referring to longitudinal differences.

While cities, jewelers, and railroads around the country subscribed to Western Union's time service in order to set their clocks and hour's bells, other entities collaborated in efforts to replicate the telegraph company's promotional time ball. In the spring of 1878, a time ball was installed on Boston's Equitable Life Insurance building across from the Old South Meeting House and several blocks from the harbor. Harvard College Observatory sent the time signal gratis, the US Army Signal Service operated the mechanism each day, and Equitable paid for the apparatus.[32] Over the next five years, eight more time balls were rigged across the country in places as far west as Crete, Nebraska; as far north as St. Paul, Minnesota; and back east again in New Haven and Hartford, Connecticut. Civic leaders hoped that the time ball standing 140 feet above street level in Kansas City, Missouri, paid for with an appropriation from the city council, would "be a prompter of punctuality." In 1880, the Connecticut Telephone Company paid to have a time ball dropped on the roof of its Hartford offices each day, unwittingly foreshadowing the time services the telephone would provide for much of the twentieth century across a good part of the world.[33] Time balls, brought to life by telegraphic transmission of time signals, presented a beguiling reference point for a culture and society developing a dependence on clock time.

As the plans for the introduction of one standard of time gained momentum, time balls figured into them as well. In 1874, the astronomer Cleveland Abbe, chief of the US Army Signal Service's fledgling weather bureau, was dismayed that he could not ascertain anything scientific about the aurora borealis from the hundreds of observations of the phenomenon made throughout the country that spring because their time stamps were unsynchronized; that is, the errors and the standards of the observers' clocks were unknown. To address this problem, Abbe had himself appointed the chairman of the American Metrological Society's Committee on Standard Time, in whose capacity he authored, had published, and circulated a proposal he titled *Report on Standard Time* (1879). This brief laid out the parameters for the system that would establish uniform timekeeping across the United States: it advocated for the North American adoption of five time standards an hour apart

Boston's time ball, 1881, from Winslow Upton, "Information Relative to the Construction and Maintenance of Time Balls," *Professional Papers of the Signal Service*, vol. 5 (Washington, DC: GPO, 1881). This remarkable photograph showcases Boston's first time ball, installed atop the Equitable Life Insurance Company's headquarters in 1878.

using the 60th, 75th, 90th, 105th, and 120th meridians, with Greenwich as the prime meridian. Abbe suggested that the best way to achieve what he called "Railroad and Telegraph Time" was to persuade railroad, telegraph, and government officials to implement it.[34] The report circulated among railway, telegraph, and government officials, who by and large greeted its proposal enthusiastically. Indeed, the chief of the US Army Signal Office assured the American Metrological Society's president in 1881 that his department wished to "contribute toward the public dissemination of 'standard time'," as it was already doing in Boston "in the maintenance of a public 'Time Ball.'" The army official stated that he was "ready to make similar arrangements with any other community that may desire it." An official memorandum listed fifty-five "stations at which the Signal Service can at present maintain time balls."[35]

Recruited by the savvy Abbe to implement standard time, William F. Allen enlisted time balls in his campaign to persuade railroads to abandon their own time standards in favor of the system of uniform standards. After receiving a copy of Abbe's report and an invitation to join the American Metrological Society's Committee on Standard Time late in 1881, Allen used his editorial position with the *Official Travelers' Guide* to publicize the need for uniform time. In April 1883, the General Time Convention and the Southern Time Convention, organizations of railroad general managers and superintendents responsible for operations, instructed Allen, their secretary, to secure the co-operation of all the nation's railroads in implementing a system of standard time.[36] Allen commenced an energetic campaign of letter writing, arm twisting, and personal visits. By August 1883, it was clear that railroads using Boston and New York time were content with the status quo, largely because their cities' prominent clocks, time services, and time balls kept their trains and customers synchronized. They followed a system of local time standards that worked fairly well.

Apparently railroad managers were worried about whether observatory time services would provide meridian time. The sense among Boston rail-roaders, for instance, was that the Harvard Observatory would stay with local time. As Allen put it, "There is some difficulty in securing the acquiescence of the roads in the vicinity of Boston," unless the time ball "can dropped upon the 75th meridian time."[37] Like the general superintendent of the Boston and Albany Railroad, regional railroad superintendents wanted "the time as furnished by the observatories [to] agree with railroad time." Allen reassured them that "the time balls at various points" in Boston, New York, and elsewhere would "be regulated in accordance with the new standards." Indeed this would happen, according to Allen, "upon the same day that the new standard [went] into effect." Allen was sure that "all of the New England railway companies [would] gladly conform to the proposed system" once arrangements were made for the time ball to drop according to the new standard.[38] So he proceeded to lobby anyone who had influence over the Harvard Observatory, which was where the time signal for the Boston time ball originated. Foremost was Professor Edward Pickering, its director, who was out of the country that summer.[39] When Pickering finally returned from Europe the second week of October, he agreed to consider the change, which in turn persuaded "Boston city authorities," to take up the question of "adopting the new standard for the city." In early November, just a week before the railroads would inaugurate standard time, the Harvard Observatory and Boston City Council resolved to adopt 75th-meridian time. A correspondent explained to Allen the source of Boston's resistance: "The old regime was so satisfactory to *her*."[40]

Allen also appealed to the keeper of the New York City time ball, the Western Union Telegraph Company. After a morning meeting in which discussion concerned dropping the time ball according to the time of the 75th meridian, which would only be four minutes later than New York City's local time, he wrote Western Union's vice president and general manager a lengthy entreaty. With Western Union's "assurance that this [could] be done in New York City, and at other points where [it] maintain[ed] time balls," Allen claimed, "adoption of the plan proposed" would be secured nationwide. Two weeks later Western Union officials had not made up their minds, so Allen asked the general manager of the Pennsylvania Railroad to weigh in on the matter. Allen's pitch to him concluded, without much evidence, that "it [would] benefit all alike to have the time ball dropped on 'Standard Time.'"[41]

Leaving no stone unturned, Allen appealed to the general superintendents of several New York railroads to pressure Western Union into recalibrating the signal that dropped their time ball. Allen also wrote directly to James Hamblet, manager of Western Union's time service, with a grandiose promise: "The time is not far distant, when the Western Union time balls in your charge, throughout the whole of this eastern district, will be dropped upon the 75th meridian time."[42] What he did not seem to know was that Western Union only dropped one time ball, the one atop its headquarters! Allen even appealed to the mayor of New York, urging him to make New York City time that of the 75th meridian, because scientific organizations and 78,000 miles of railroads depended on it. What is more, he argued, since Boston had adopted 75th-meridian time, New York needed to claim it too, lest the new railroad time be synonymous with "Boston time."[43]

Time balls seemed to be Allen's solution to the challenge of disseminating the new time standards of the 75th, 90th, 105th, and 120th meridians. A time ball in San Francisco was the technological fix Allen proposed to the general superintendent of the Central Pacific Railroad, who still had not agreed to the new plan for standard time with fewer than six weeks until the proposed adoption date. After noting that the difference between 120th-meridian time and San Francisco's local time was but ten minutes, Allen reasoned, "It would seem as if you could readily conform to the proposed system." He then pleaded, "If a time ball is dropped in San Francisco, and it could be dropped on the new standard, people would almost immediately set their watches by it." Grasping at straws, Allen noted that the time ball would be a boon for navigators too. If the Central Pacific railroad would adopt 120th-meridian time, "exactly eight hours from Greenwich time," Allen forecasted, the Pacific coast would become synchronized with Greenwich. Western values would reign.[44]

Few cities had time balls, so it was their most prominent public clocks that were enlisted in the cause of banishing local time and crowning standard

San Francisco's time ball, 1884, Roy D. Graves Pictorial Collection, San Francisco Views, vol. 3, the Bancroft Library, University of California, Berkeley. This Pacific coast time ball was hoisted atop the Telegraph Hill's Pioneer Park Observatory. It may have received telegraphed time signals from the Mare Island Navy Yard Observatory.

time. Most cities and towns acquiesced, agreeing to recalibrate their public clocks and bells to the new time standard. In Philadelphia, for instance, before the change, questions were asked about the deviation between meridian time and local time as shown on the Old State House clock.[45] On the day the new system began, the city of Philadelphia and the Pennsylvania Railroad adopted it, and according to Allen, "the clock on Independence Hall was changed thirty six seconds," which brought it in accord with 75th-meridian time.[46]

The US Naval Observatory was also a willing partner in implementing a national standard time. In response to a letter from Allen requesting a favorable outcome to Western Union's request for the observatory to "furnish them with the time of mean noon, 75th meridian time," the observatory's superintendent congratulated Allen on the plans for standard time. Not only did the superintendent agree to furnish 75th-meridian time to New York, but he also promised to "try to secure the immediate adoption of the same time, as the local time for the whole section in which it is to be used by the Railroads." Rear Admiral Shufeldt explained that the observatory was "furnished with automatic apparatus for sending time by telegraph, and does so send it each day over the lines of the Western Union Company."[47] Assigned the responsibility for implementing a system in which standard time signals would be sent, another observatory official explained to Allen that "one signal over the wires" would be sent to railroads. This same signal, it was promised, would drop time balls in New York, Philadelphia (where a time ball was not yet installed), and Washington. The observatory official thought the railroads' plan to follow one standard of time "would act as an inducement to the public to adopt the same as their local time," which he considered "an important part of this scheme."[48]

The dogged Allen and nearly all the US Naval Observatory staff agreed about the need to dispense with local times. In 1881, the observatory had spearheaded a failed plan to abolish the District of Columbia's local time.[49] As the plans were coming together to implement standard railroad time, Allen became increasingly frustrated with the schemes put forward meant to simplify the translation of this new time into a town's true time. He estimated that "there are probably not over a dozen cities in the United States that have any means of finding out what their *true local time* really is." Allen derided this sort of calculation as "simply a snare and a delusion."[50] Wearing his editor's hat, Allen asked in the September 1883 issue of the *Travelers' Official Railway Guide*: "Can any one tell us precisely what local time is really?" A litany of examples of the loose timekeeping habits of most American communities followed. Allen mused in anticipation to one of the US Naval Observatory's astronomers, "'Standard Time' will soon be used so entirely that it will be unnecessary to refer to local time for any practical purpose." He further exhorted, "Local time is practically unknown in Great Britain to-day," because, "the people conformed to the new standard" when it was introduced in 1848. Standard time took hold there, Allen explained, "without the aid of such powerful monitors as time balls." With the assistance of time balls, he argued, "I do not think there can be a shade of doubt that the people would conform almost, if not immediately to the new standard." Arrogantly he predicted that "if the watches and clocks were changed, without the people knowing it, not one out of ten thousand would find it out for himself."[51]

On Sunday, November 18, 1883, as Allen and his wife and sons stood on the roof of the Western Union Telegraph Company building around midday, they "heard the bells of St. Paul's church strike the old time," watched the Western Union time ball drop as it heralded the debut of a national standard, and rejoiced as Trinity Church's bells announced the new time.[52] Standard time established that it was the top of the hour everywhere at the same moment: four regions divided at longitudes fifteen degrees apart were set to four different hours moving east to west.[53] Not everyone in the large nation saw the time balls drop at the new noon on Sunday; indeed, with only eight operational time balls in the United States, public clocks and bells like those on St. Paul's and Trinity did nearly all the work of distributing the new standard time. On the following Monday in Chicago, "knots of people" were "found grouped about every jeweler's window where the standard time is displayed, anxious to set their watches by the new time."[54] US Naval Observatory time, as Allen predicted, did gain preeminence under the new system. For instance, promotional copy for the newly formed Time Telegraph Company emphasized how "the pendulum of a central regulating clock, which stands in the [Time Telegraph] company's office, runs exactly according to the time as furnished from the National Observatory in Washington."[55] Now rather than standards of time that changed from place to place, there was a standard time that was impervious to all geographic indices, except that of the artificially constructed boundaries of time zones.

In most places, standard time did not diverge dramatically from the rhythms of natural time to which people were accustomed, but communities on the edges of time zones initially balked at the new procedure. In those places, standard time was more than thirty minutes fast or slow of local time. Some of these communities dug in their heels, determined to keep their local time. Just a few days before the introduction of standard time in 1883, Cleveland's wholesale jewelers Brunner Brothers ordered a jeweled No. 61 E. Howard regulator with a red minute hand "set 28 minutes slow, to designate railroad time."[56] The next year, E. Howard received more orders for clocks with multiple minute hands from places near boundaries between time zones, including one from a Kentucky courthouse that wished to have red minute hands announcing local time and gilt ones showing standard time.[57] Over the decades, adjustments were made to the boundaries of the time zones, and communities gradually let go of their fealty to local time. The substitution of national standard time for local standards of time exemplifies the transition from what political scientist James C. Scott identifies as "irreducibly local measurement practices" to ones that are "legible" and thus especially useful to national states seeking to establish control over "a far-flung, polyglot empire."[58] The new national time's legibility depended most heavily on public clocks, multitudes of them, consistently announcing standard time.

Admiral John Rodgers, once superintendent of Washington's observatory, opposed standard time in the early 1880s. His solution to the excesses of local time had been to provide all clocks in railroad depots with two pairs of hands, black hands for local time, and gilt ones for Washington time.[59] His design for gilt hands showing Washington time returns us to John W. Bell's patented coupled pairs of watch hands that could keep two times. Bell meant for one pair of hands to show a standard time that all railroads in the nation would follow, the other pair to indicate local time. If all watches were equipped with them, then, according to Bell, "no person would be inconvenienced by the establishment of a single standard time." A traveler wishing to use the railroad could consult a timetable and the two sets of hands on his watch. If he was on an ocean voyage, he could set one pair to his home's meridian time, and then "each day when ship time [was] changed," he could reset the other hands and "see the space widening between the [pairs of] hands." Bell was particularly enchanted by this idea: "The passenger can thus tell how far he has come & how far he has yet to go in a way that is both agreeable & useful." And once abroad, the traveler could take "comfort [in] this keeping your home time while in a strange land."[60] For Bell, the coupled pairs of watch hands were not only useful, but also a source of pleasure and solace.

Shortly after receiving his patent in November 1882, John Bell appealed to William Allen to spread the word about the utilitarian possibilities the coupled pairs of watch hands offered. Others had made similar appeals to Allen, including an engineer on the Southern Pacific Railroad who designed, but did not patent, a watch with two sets of hands that could show standard and local time simultaneously.[61] Early on, Allen had repudiated such commitment to local time, particularly those expressed in rival Charles Dowd's self-published and widely distributed time difference charts of the 1870s. These tables showed the difference between railroad time and local time for eight hundred railroad stations. Dowd, a schoolteacher, was determined to ameliorate the confusion caused by what he called the "bewildering diversity of clocks in the same depot." In 1877, he floated in print the idea that railroads adopt a standard time set "fifteen degrees Longitude apart." The next year, Dowd published another pamphlet, *The Traveler's Railway Time Adjuster*, which he hoped would "enable anyone to keep his own time, in whatever part of the country he may be traveling." Ignored over the years, Dowd finally penned his own history in the 1884 pamphlet *System of Time Standards*, where he recounted presenting the time zone system to a convention of railroad managers in 1869 and asserted petulantly, "Standard Time is not precise Local time."[62] But who cared any longer? Allen, who held no affection for or attachment to local time, had for all intents and purposes abolished local time.

It is therefore decidedly odd that in 1882 Allen allowed Bell to add an extra pair of hands to his watch. Stranger still, Allen called it "a convenience" in a letter to Philadelphia and Chicago jewelers urging them to offer their customers the service of adding extra watch hands.[63] Although Allen's voluminous and at times revelatory correspondence does not indicate what he was up to when he carried and even promoted Bell's watch showing multiple times, it is clear that the device, just like time balls, was part of his strategy for winning over railroad men to the standard time plan. Bell's technological fix may have held more appeal for Allen than Dowd's complex time difference equations and charts, but more important, Bell did not seek to deflect attention away from Allen as had the persistent Dowd. Perhaps Bell's wealth, connections, and generosity also appealed to the ambitions of Allen.

So despite his disdain for local time, Allen was a good mark for the promotion of Bell's paired watch hands, what with his leadership of the effort to induce American and Canadian railroads to adopt a single standard of time. Indeed, for the two April 1883 railroad conventions (the General Convention and the Southern Convention) at which Allen planned to discuss "the Standard Time question," he asked Bell to supply him with "a large silver watch with double hands on it to pass around." Allen approved of the stem-winder watch Bell sent, particularly "the fact that the hands are distinctive enough in appearance." At both conventions, at which it was provisionally agreed to adopt standard times, the watch was "extensively handled," but Allen was disappointed to report to Bell that "there was somewhat less interest in it than [he had] anticipated." He hoped that there would be more after standards were "put in force."[64] Allen's hopes were misplaced, or perhaps disingenuous.

The lackluster interest in Bell's innovation arose out of a miscalculation on Bell's part, that is, that people wished to keep two separate times simultaneously. Bell pointed out that sailors had been doing so for centuries, and landlubbers had been since the 1830s and 1840s. He thought people "not so blind or stupid that they must needs mistake one [time] for the other," but instead saw them as mathematically challenged when it came to calculating time differences and translating one time into another. As Bell lamented in an unpublished manuscript penned before the adoption of a national standard time: "We have perfected our watches until they may be said to run without variation or error for all practical purposes, but with all this accuracy they become tormenting puzzles as soon as we get a few miles from home." The tired traveler wishing to catch a train "goes groping about with a watch & a time table in his pocket which though both are perfectly accurate are at times almost useless to him." In only a few places did the local time correspond with railroad time. So when "a New Yorker goes to Washington he travels by four different times," which as Bell pointed out, "is only one instance of hundreds all over the country where we have 58 different standards of time for our Rail Roads."[65] Another tinkerer fascinated by dual time standards was the young Henry Ford, who held an abiding interest in watchmaking. In 1883, he put together a watch with two dials, one showing local time, the other railroad time.[66] Around the United States in the early 1880s, the multiplicity of times a clock could show challenged tinkerers like Dowd, Allen, Ford, and Bell. Allen and Ford moved toward what would define the future—standardization of time and industrial production—but Dowd and Bell were stuck on multiple times.

An encounter Bell had with an old German watchmaker underscores just how out of step he was with his contemporaries. After making what Bell described as "one very good set," the watchmaker declined the commission to make as many coupled pairs of hands as might be ordered. As Bell was paying him for the work, the watchmaker looked at him "in a comical way," asking, "What is the object[?] What use is it[?] If I want to calculate a difference of time[,] I can do it near enough." Bell, taken aback, responded, "that is just the point[,] it is not near enough for this age." He then reminded the watchmaker, "There was a time when watches only had one hand, an hour hand, people could calculate near enough for the minutes." The old watchmaker agreed, "Don't I know that." He reflected, "I well remember the first watch I saw with a minute hand, it was so puzzling, *it was a long while before I learned to tell time with it.*" Bell took heart from this exchange, figuring it would be a long while before people could tell the time with two pairs of hands on the watch face, but that it would provide the exactitude and certitude necessary for the age.[67] Instead, in the public clock era, accuracy would be achieved by implementing a single standard of time, dispensing altogether with the variegated local times that were so troublesome.

Despite the lack of interest among railroaders and the German watch-maker's skepticism, some demand for Bell's coupled watch hands developed. However, initially no watch company would agree to make the hands, so Bell could not supply samples or price lists to the watchmakers who sent requests from, as he characterized it, "every quarter north, east, south & west." He decided to go it alone, as he put it, "to carry the watch into Africa." His Africa was Boston. On a winter's visit there in 1883, months before Allen's triumph, Bell toured the Waltham Watch Company's factory, which he called "one of the wonders of the world," but he was not able to persuade its own-ers to make his patented watch hands.[68] No doubt Bell's distribution of a leaf-let to railway officials across the country, as well as an advertisement he ran in the *Official Railway Guide*, generated some business in the early months of 1883. With an engraving of two watches, one "with supplementary hands visible," and a caption reading "IMPORTANT TO WATCH BUYERS," Bell's advertisement cautioned: "In view of the fact that Railroad standards of time have been reduced nearly one-half in the approach toward a single standard, . . . no one should buy a watch that cannot keep double time at the pleasure of the wearer."[69] Along with a check for $27.50 to pay for running the notice for three months, Bell explained to *Guide*'s editor (Allen) that he considered "the money as spent not invested." He believed that offering a watch with two sets of hands would help "to bring about the adoption of standard time."[70] Perhaps Allen believed this too.

Bell, like many others in the early 1880s, maintained fealty to a local time that approximated sun time. In his view, its uniqueness fostered local pride. Like Admiral Rodgers, Bell felt that the sun should regulate all life and thus should be the local standard of time.[71] Neither one of these men anticipated how the extension of electricity and electric light into public spaces and pri-vate homes, which began in the 1880s, would eclipse the primacy of sun time.[72] What does the sun matter when illumination can be perpetual? What is more, as telegraphy extended its reach and a telephone network developed, many other entities besides railroads were developing widespread connec-tions that demanded coordinated time standards. So Bell's innovation was less a forward-looking technological fix to the multiplicity of time standards than an expression of stale assumptions concerning mechanical time. To wit, the patented watch hands assume that mechanical timekeepers ought to show sun time, which ought to remain preeminent because the sun "regulates all life upon earth" (as Admiral Rodgers put it).[73] Behind Bell's patent was the sense that only railroads needed standard time because only they intersected with local times at a fast enough pace to warrant such measures. But Bell was wrong. As the pace of life quickened, a dizzying array of timetables joined that of the railroads; one standard of time eased the task of coordination.

By the 1890s, nearly all American communities—Detroit being a notable exception—followed a national time standard conveyed by authoritative public clocks. The easy acceptance of the arbitrariness of the new system was, as Peter Galison puts it, "as great a transformation as the acquisition of a regularized time awareness."[74] Standard time had become a convention, a procedure, a conceit. Birthed in 1883 as railroad time, it became national time within few years.

CHAPTER 5

American Synchronicity

Turn-of-the-Century Tower Clocks, Street Clocks, and Time Balls

Nine out of ten are ready to swear by the City Hall clock. . . . At the first stroke of noon, out come the watches, gold, silver and nickel and when adjustment is requested it is at once and unhesitatingly applied. . . . Perhaps the majority rely upon the old clock implicitly, and like to be hand in hand with it, as it were, experiencing a feeling of deep gratification whenever perfect accord exists between it and their timepieces.

— "Four Faced, But Honest," *Brooklyn Eagle* (1890)

After the nation began to follow standard time, Americans living in towns and cities depended on public clocks for a sense of synchronicity, or what one period journalist called "perfect accord." City newspapers featuring articles with witty titles like "Four Faced, But Honest" extolled the virtues of their town clock.[1] Public clocks inspired devotion, but they also elicited occasional mistrust, as reflected in newspaper article titles like "No Tick There" or "What Time Is It?"[2] Again and again, even after 1883, it was remarked how "rarely are two clocks found registering exactly the same" and how "almost impossible to judge which of a number of clocks is correct."[3] The multiplicity of clocks, each keeping its own time, was confusing, but some garnered more attention and trust than others, and none more so than government clocks. Governing authorities in the United States invested more heavily in public clocks than did the central agents of modernization, factories and railroads. City, county, state, and national government paid for one-third of all new public clocks between 1871 and 1911 (see table 2). While Gilded Age city bosses and national political leaders often showed themselves to be duplicitous and untrustworthy, their clocks, even when unsynchronized, accrued affection and devotion.

To be sure, other clocks exerted powerful pulls on habits and imaginations, none more so than those belonging to railroads, despite the fact that as a whole they accounted for fewer than 6 percent of all new public clocks in the United States between 1871 and 1911 (see table 2). During the decade prior to the implementation of standard time, when terminals and depots were often sited at the far edges of towns, only one out of ten new railroad clocks was installed on or near the building's exterior. But between 1884 and 1890, railroads installed at least a third of their clocks on the exterior of their depots, with most of them rising above cities atop towers. Improvements in tunnel engineering, elevating railroads, and harnessing electricity in the 1880s and 1890s brought some railroads and their depots to more central locations. The grandest railroads serving the largest cities built monumental passenger stations as close to city centers as possible.[4] In 1881, Philadelphia's Pennsylvania Railroad Passenger Station, known as the Broad Street Station, with its imposing neo-Gothic brick clock tower, was completed near the city's newly burgeoning commercial district.[5] As skylines reached higher, some railroads installed street clocks outside their terminals, attempting to capture pedestrian attention. Philadelphia's Reading Terminal, opened in 1891, distinguished itself with a fifteen-foot-high four-dial post clock that simply announced "Reading Terminal," which complemented the enormous street clocks of jewelers on nearby Sansom Street.[6] Bundled into a railroad clock were a set of powerful meanings for clock time associated with speed, efficiency, accuracy, power, and standard time itself. Railroad clocks and railroad time pushed Americans toward modern time discipline.

While most of the nation may have accepted railroad time as the standard, it was largely civic and commercial clocks that made this time a material reality. Church and school clocks did their part too, particularly because their bells reached beyond their congregations and classrooms. Aurally and visually these timekeepers presented clock time as impartial, as transcendent, and as an American patrimony accessible to all. As elsewhere, in the United States timekeeping bolstered state authority and legitimacy.

I begin this chapter by highlighting some examples of the use of clocks to assert political authority from world history. I then look closely at the United

Cervin Robinson, *Concourse from South*, Pennsylvania Station, New York, April 24, 1962, HABS NY, 31-NEYO,78–79, Prints and Photographs Division, Library of Congress. Pennsylvania Station, home to an array of stunning public clocks, opened near Manhattan's center in 1910.

States, where after the Civil War, municipal and commercial entities invested in clocks to enhance their authority. Within this story, I outline the efforts of jewelers to sustain their temporal authority in an era when spectacular, large, and unusual clocks overshadowed their regulators. To close the chapter, I consider how the federal government systematically invested in clocks, time balls, and timekeeping systems that enhanced its physical and symbolic presence throughout the nation. Federal agencies did not succeed in doing this alone, but instead worked in partnership with public and private organizations. While my discussion shows that far from perfect accord characterizes the public clock era, it demonstrates how the ideal of synchronicity inspired city councilmen, employees of branches of the US government, small shopkeepers, corporate architects, jewelers, and inventors.

A brief survey highlights the centrality of timekeeping, which connotes authority, permanence, and stability, to governance across space and time.[7] Dating back centuries, ruling authorities in the East and West have claimed the responsibility and right to keep and tell the time. In China, as art historian Wu Hung explains, "various kinds of advanced chronographs, including some extremely complex hydraulic clocks, were invented under imperial patronage to serve the symbolic role of legitimating political authority." Hung describes a seventy-foot-high water clock built in 1262 next to the Imperial Bright Hall, where the emperor acted "as the moving hand of a huge clock" when he moved from room to room in accord with the months. Several centuries later, an eighteen-foot-high Western-style clock was placed in the Hall of Union next to the imperial throne, which was flanked on the other side by a traditional Chinese water clock. The emperor sat between these gigantic clocks, as Hung points out, "for symbolic not practical reasons." Monumental timekeepers bolstered claims to power and authority within the imperial palace, while across China public timekeepers were under imperial control. Between the third century BC and the earliest years of the twentieth century, imperial edicts mandated the beating of large drums or ringing of large bells at the time of the opening and closing of the gates of cities and towns.[8]

In early modern Europe, public clocks were also instruments of the ruling class. Territorial rulers like Edward III, Charles V, and the dukes of Burgundy provided for their erection directly with subventions and indirectly through levying taxes; they also punished communal rebellions with the removal of the village clock or bell.[9] The Great Clock of the Palace of Westminster, whose tower is home to the bell known as "Big Ben," was installed in 1859, with the prayer "O Lord, keep safe our Queen Victoria the First" inscribed in Latin on each of its four enormous glass dials. The Houses of Parliament's timepiece, whose architect planned for it to be a "King of Clocks," asserted dominion over the British Empire. Today it is an icon for Great Britain, for London in particular.[10]

Around the world in the nineteenth and early twentieth centuries, clock time belonged to a contingent of strategies states used to stabilize and control societies. In the 1820s and 1830s, colonial officials in Australia leaned heavily on clocks to establish authority in and beyond the penal colonies that had brought Europeans to the continent; schedules, bells, and other timekeeping regulations emphasized discipline and control.[11] Japan's national government also deployed clocks in its effort to modernize economic and social organization. It adopted Western time in 1873; several years later the Colonization Commission charged with modernizing the northern island of Hokkaido and its principal city, Sapporo, funded the construction of a twenty-foot-high clock tower, the Tokeidai.[12] During the same era, the Ottoman sultan Abdülhamid II initiated several programs for the construction of clock towers meant to, in the sultan's words, "show western time and automatically chime the hours." In honor of his silver jubilee in 1901, Abdülhamid called for each provincial capital to build a clock tower in his name.[13] After the Chinese Republican Revolution of 1911, political leaders appropriated government revenues to build and equip prominent clock towers, as did revolutionary regimes elsewhere in the early twentieth century.[14]

The association of clocks with political authority extends to the United States. Their installation and maintenance constituted local and national political acts. Through the middle decades of the nineteenth century, government clocks, like the buildings upon whose exteriors they were found, were small affairs.[15] After the Civil War, local and county government appropriated monies for multiple exterior and interior clocks on their administrative buildings, including town halls, county courthouses, fire departments, and police headquarters, which were often housed in the same building. Large clocks and hour's bells were installed upon city halls and courthouses, each a signal space in the United States, as historian Mary Ryan argues, "in which to imagine and to practice representative government."[16] Unusual was a public clock on a state capitol, perhaps owing to the repetitive use of a set of authorizing symbols derived from classical architecture (the dome, rotunda, and portico).[17] Until the end of the century, public clocks were not much in evidence in the exterior public spaces of the nation's capital, though they did tick away inside the halls of federal government and in the many federal buildings constructed around the nation.

Mechanical timepieces were also important to the symbolic conceptualization of political power and national destiny in the United States, as I noted in the discussion of "republican heirlooms" in chapter 3. Literary historian

Thomas Allen explains that clocks and clock time provided a platform upon which antebellum American political thinkers dreamt of "an empire extending through time rather than across space." He shows that Thomas Jefferson, the president who made the Louisiana Purchase, and John O'Sullivan, the architect of "Manifest Destiny," were obsessed with the conquest of time. American intellectuals, Allen also points out, used clocks and clock time "to imagine different forms of capitalist market exchange and modern national identity."[18] Like books in Renaissance portraiture, watches and clocks visually conveyed power, authority, and legitimacy to those in proximity. Some of the photographer Matthew Brady's most politically ambitious clients chose to pose with a mantel clock for their portraits: General Robert E. Lee, President James Garfield, and Red Cross founder Clara Barton, for example, chose the clock from among props that included a thick book and formal columns. Brady's studio clock, marketed by the Ansonia Clock Company as "the Reaper," featured a brass figure of a youth with a sheaf of wheat. On the clock dial was engraved "M. B. Brady, Washington, D.C." In extant photographs, the clock's hands always show the same time, 11:52.[19] Adorning one's space, one's body, one's city, or one's nation with timepieces was as assertive a political act as that of the Chinese emperor who sat on a throne between towering clocks.

City authorities made noteworthy responses to the introduction of standard time. By April 1884, nearly all the major American cities had made standard time their official time. With the exception of Ohio and Pennsylvania, no more than two towns in any state retained local time.[20] While many cities officially adopted standard time, only a handful of state legislatures confirmed it. Some governors commended standard time, urging localities to adopt it, but most felt it was beyond the purview of state power to set the time. The lack of attention at the state level to time standards reflects a well-developed pattern since the Revolutionary era. Local authorities were all too happy to maintain their historic responsibility for the time, even as the federal government assumed sovereignty over clock time and installed national clocks and time balls wherever it could.

Jewelers persisted in their assertions of temporal authority during these decades. After the railroader William F. Allen's triumph in nationalizing time, many jewelers obtained clock dials, glass doors, or cabinetry lettered with "standard time." A majority of such orders placed with E. Howard and Seth Thomas were from shops in southern and western cities, like Topeka, Dallas, Little Rock, and Raleigh, all places far from observatories and telegraph time services.[21] A few weeks in advance of the introduction of standard time,

Matthew Brady, portrait of General John A. Campbell, 1863, Matthew Brady Photographs of Civil War–Era Personalities and Scenes, Record Group 111, National Archives Records Administration. After the Civil War, General Campbell was appointed first governor of Wyoming Territory. He is pictured here with a clock in Brady's Washington, DC, studio. Like many other Americans who aspired to power, General Campbell chose a mechanical timepiece as a status object.

the New Orleans Produce Exchange ordered an elaborate three-dial gallery clock from E. Howard. The top half of the twenty-four-inch center dial was marked "M. Scooler," the bottom half "New Orleans." One sixteen-inch dial was marked "San Francisco," the other "New York." In the same order placed with E. Howard, the Canal Street jeweler whose name is at the center of this arrangement—Maurice Scooler—also requested a clock for the

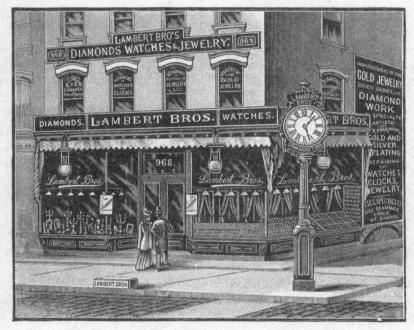

LAMBERT BROS.,

Third Avenue, cor. 58th Street,
NEW YORK,

Manufacturing ꞓ Jewelers ꞓ and ꞓ Importers.

The Most Popular Jewelry House in New York.

The PIONEERS and ORIGINATORS of selling———

STANDARD QUALITY GOODS

———at MANUFACTURERS and IMPORTERS prices.

"Standard Time" clock in front of Lambert Brothers, 1895, Baker Library Historical Collections, Harvard Business School. This advertisement shows a busy New York watch and clock shop with a street clock outside engraved "Standard Time." Lambert's ordered a post clock like this one from E. Howard on April 4, 1888.

New Orleans Sugar Exchange. The specifications for this one indicated that a tower clock movement would be connected to one four-foot illuminated dial engraved "Scooler's Time."[22] Others branded the time that their clock meted out as "city time." A Los Angeles jeweler ordered a clock in 1884 with the dial glass painted "Frederick Linde, City Time." The Nordman Brothers of Walla Walla, Washington, ordered their clock with "City & Rail Road Time" painted on the pendulum's glass door. Still other jewelers held fast to the title city timekeeper. In 1890, Salt Lake City's firm Davidson, Leyson and McCune asked that its No. 75 clock be lettered with their names and "Authorized City Time Keepers."[23]

Jewelers likely owned 10 percent of all new public clocks installed in American cities and towns between 1871 and 1911 (see table 2).[24] Most did not occupy the kind of real estate hospitable to clock towers, so instead they moved their clocks out of windows and interiors into the streets. Post clocks, also called street clocks, were a bold way for jewelers to reassert their claim to authority over public time in the decades during which railroads, governing agencies, and churches built ever larger clocks and hung ever heavier bells. Extensive advertising in local newspapers shored up the promotional efforts jewelers expended in the installation and maintenance of these costly timepieces. The most elaborate post clocks, such as the one built in 1907 for Jessop's jewelry store in San Diego, ran multiple dials and stood as high as twenty-two feet.[25]

Not all jewelers had the means or skills to install street clocks, which were expensive to construct, purchase, and operate. Oftentimes, proprietors simply used the image of a clock or watch dial to distinguish their signage. African American jeweler E. J. Crane, who claimed to be "the only Colored man in the South ever to make a watch out and out," dangled a double-dial clock in front of his turn-of-the-century shop in Richmond, Virginia. He advertised extensively in the local African American newspaper, the *Richmond Planet*.[26] A photograph of New York City's Mulberry Street taken sometime during the first decade of the twentieth century shows two large pocket-watch-shaped signs—one announcing "Ingersoll Dollar Watch," the other the street address—hanging from an awning that jutted out over an endless line of produce peddlers.[27] Regardless of their resources, jewelers at the turn of the century relied on clocks perched above crowded streets to draw attention to their operations and to stake a special place for themselves as authorities for the time, rather than just merchants of timepieces.

What with various local establishments including jewelers, churches, banks, insurance companies, and large stores, installing and maintaining public clocks, the efforts of local governing agencies to disseminate the time to the public take on added significance. Always short of revenue, often having

Storefront of Philadelphia jeweler Aug. Gehring, ca. 1880, photograph by A. W. Rothen-
gatter and Company, courtesy of the Library Company of Philadelphia. An early version
of the street clock was a double dial affixed to a post. The tubing through which the dials
were attached to a clock movement inside the shop is visible in this photograph.

San Diego History Center

Jessop's jewelry store, San Diego, 1910, San Diego History Center. Jessop's street clock had four large dials, along with sixteen smaller ones showing the time of cities around California and the world. It won a gold medal at the 1907 California State Fair before being installed in front of the San Diego jewelry store, where it was not only a reliable timekeeper, but also a tourist attraction. Today it stands in San Diego's Horton Plaza.

Storefront of watchmaker E. J. Crane, Richmond, Virginia, 1899, LC-USZ62-118028, Prints and Photographs Division, Library of Congress. This photograph was displayed as part of the American Negro Exhibit at the Paris Exposition of 1900. Crane, who began repairing watches in 1875, was a prominent member of Richmond's African American community after moving to Virginia from Georgia in 1892. Not only did he repair watches and jewelry in his Broad Street shop, he also offered for sale pistols, watches, watchcases, and jewelry.

to appeal to parsimonious, antiurban state legislatures for funds, it is surprising that 61 percent of the towns and cities across the nation doing business with E. Howard and Seth Thomas purchased tower clocks, rather than less expensive timepieces, or none at all. As "monuments to democracy," city halls made their claims for authority by activating different mixes of authoritarian, confrontational, and communal architectural strategies.[28] During this period, when more city halls were built than at any other time in American history, the clock was a central motif, found on their domed towers, cupolas, steeples, and precariously tall towers. Particularly grandiose clocks were installed on the Louisville City Hall (1870–73), the Philadelphia Public Buildings (1870–1901), the Minneapolis City Hall and Courthouse (1889–1905), the City and County Building in Salt Lake City (1892–94), the Milwaukee City Hall (1893–95), the Savannah City Hall (1904–5), and the Oakland

Milwaukee City Hall, 1904, WHi-48057, Wisconsin Historical Society. In 1896, when a tower clock was installed in the Milwaukee City Hall, it had the biggest clock face in the United States, with a diameter of fifteen feet. The large, heavy hands were so far from the clock room that a unique system using pneumatic tubes moved them. Warren S. Johnson, the same entrepreneur who would a few years later sell the city of Philadelphia its troublesome clock, designed it. The bell, weighing twenty thousand pounds, was one of the heaviest in the world: it took sixteen hours to hoist it into the tower.

City Hall (1911–14).[29] In addition to city halls, clocks adorned other civic buildings, including public high schools, libraries, police stations, firehouses, market houses, and ferries.

Though many smaller towns did not have a designated city hall or other city building worthy of a tower clock, they also obtained large public clocks. In some instances, city officials shared quarters with county commissioners and judges; together they funded buildings housing both county courthouses and city halls.[30] In other instances, town councils appropriated funds to erect a clock on their locality's most prominent tower, which typically belonged to a church. For example, the new town clock for Northampton, Massachusetts, which was intended to provide "an authoritative source of time," was installed in 1878 in the First Church's tower.[31] Enthusiasts for projects like this sometimes made vaulting claims: "A church tower without a clock and bells seems like an unfinished edifice . . . like a form without life, a body without soul." A clock, however, was not simply an architectural detail, but an essential component of an orderly community. It was said that when a "clock stops, it produces social discomfort, and anarchy throughout a whole neighborhood, to an extent scarcely credible."[32] Even when purchased with private funds, church clocks were included in the pageantry of the state. For instance, in 1893 when the Free Baptist Church of Ashland, New Hampshire, ordered a special striking tower clock for dials spanning five feet in diameter and a 1,200-pound bell, they insisted that "the clock be up and running July 4."[33]

County courthouses, found in cities and small towns, also sported large clocks and heavy bells. More than half the government clocks in the South, for instance, appeared on county courthouses, reflecting the centrality of county seats in rural areas.[34] After the Civil War, much of the remaining local pride and money was invested in the seat of county power, the place where deeds were recorded, marriage licenses issued, properties appraised, taxes levied, court decisions rendered, and estates administered. It is no wonder that the eclectic buildings, often made of local materials, featured towers, which, when the funds became available, were adorned with clocks. Whether simple or grand, the hoisting of clock faces onto a tower, cupola, dome, or pediment signified the county's ambitions toward the cosmopolitan and modern, and if not exactly that, then at the very least toward the orderly. Between 1890 and 1905, 23 of Georgia's 137 counties installed new E. Howard tower clocks; tower clocks still distinguish many courthouses dating from the 1890s and 1900s.[35] Texas county commissioners also made expenditures for courthouse tower clocks; thirty of the estimated fifty Texas county courthouse hour-striking tower clocks installed in the Gilded Age appear in the E. Howard and Seth Thomas accounts.[36] In the cities of the New South, the county courthouse often joined the post office in buildings entirely out of scale with their surroundings, such as the Nashville post office and courthouse, which sported a 190-foot clock tower, or Birmingham's Jefferson County Courthouse.[37]

In the United States, the presence of clocks within courtrooms or on the exterior of courthouses underscored the temporal power of governing authorities. Indeed, it was the courts that ultimately decided in several cases whether mean or solar time, local or standard time, or daylight saving time was the time-reckoning system in effect when an action or event (like a fire) ensued. "Because there was so little legislation spelling out the accepted way to tell time," one legal scholar explains, "numerous lawsuits in which time's reckoning was an issue occurred."[38] But a court's power over time is far more extensive than simply adjudicating the issue of time reckoning or setting out temporal parameters for court proceedings. Courts of law have historically traded on their power over the time of all social actors under their jurisdiction.[39] Judges and juries are granted the right to extract time from defendants through parole sentences, jail terms, or in some instances through the death penalty. By 1900, the deep-seated symbolic connection between the law and time was made evident in the form of a clock in nearly every courtroom and on most courthouses in America.

The contest for authority over civic space and time simmered in all American cities during the public clock era. That conflict erupted in Milwaukee during the first decade of the twentieth century, leaving the city's esteemed jewelers with junked clocks and the city with fewer sources of public time. On a March night in 1908, twelve firemen—on the mayor's orders—razed eight street clocks. Each belonged to a different jeweler; each weighed more than a thousand pounds. The firemen used hammers, saws, and chisels to remove the bolts anchoring the clocks to the sidewalks, detach the electrical circuitry, and dislodge the posts from the sidewalk cement. The clocks were destroyed in the process: as they toppled over, the glass covering their dials shattered; the brass ornamentation broke off; the clock faces themselves bent under the weight of the fall. The next day, newspapers published accounts of the nighttime raid accompanied by photographs of the leveled clocks. Milwaukee's mayor, Sherburn Becker, a progressive reformer and member of the Republican Party, classed the street clocks as hazards akin to produce stands and other forms of illegal public commerce. In his effort to clean up his city, he drew no distinction between jewelers and grocers, demanding that they all clear the sidewalks of clutter. His was a crusade against "the nuisance of signs."[40]

Menu cover for a dinner for Judge William S. Kochersperger at the Bellevue-Stratford, 1907, Buttolph Collection of Menus, Rare Books Division, New York Public Library. The connections between the law and clock time leap out from the cover of a humorous menu enumerating the drinks, foods, and guests at a celebratory gathering of Philadelphia's Columbia Club in 1907. It shows a judge whose arms are filigreed clock hands; his decision: "You're Late and Must Pay the Penalty!"

Destroyed street clock, Milwaukee, 1908, courtesy of Milwaukee County Historical Society. This was one among many photographs that dramatized the destruction of street clocks belonging to Milwaukee's jewelers.

Several days after their clocks "WERE ABSOLUTELY WRECKED," the jewelers of Milwaukee placed a paid advertisement in the city's leading English-language newspaper. They explained that when it "became known to the public" that the order to take down signs included public clocks, "the Jewelers were besieged by many of their patrons and the public in general to make every effort to retain the clocks as they were considered a public convenience." The jewelers visited the mayor in person to protest the comparison of their clocks to signs and posters.[41] In response, the mayor claimed he could not exempt street clocks from prohibitions against street advertising, because soon enough there might go up "a clock with a mug of beer or a pretzel painted on its face."[42] After reassuring the mayor that their clocks were meant as a public service disseminating the time, not as advertisements, the jewelers mistakenly thought they could "let the matter rest." But the mayor persisted with his campaign, and their street clocks were demolished. The jewelers' resulting outrage was not only about the loss of their "EXPENSIVE CLOCKS," but about the city's loss of prestige: "MILWAUKEE IS THE ONLY CITY IN THE UNITED STATES WITHOUT STREET CLOCKS."[43]

In the aftermath, the jewelers "absolutely refused to touch the damaged clocks," leaving the debris on the sidewalks for the city to clean up.[44] The Democratic political machine soon capitalized on Mayor Becker's misstep. What could better show the lack of fit for higher office than the destruction of clocks? After his own party declined to nominate him for reelection as mayor, Becker moved to New York City and there became president of a bank.[45] Archie Tegtmeyer, one of the jewelers' spokesmen, observed that most people regarded the clocks "as public utilities." He explained that in other leading midwestern cities "where signs were ordered removed from the streets," it was nevertheless the case that "the clocks of jewelers were permitted to remain." Another advocate for the jewelers argued, "Clocks are an ornament rather than a detriment." In his view, they were "no greater a menace than lamp posts."[46] A week after the midnight raid on their clocks, Milwaukee's jewelers decided to donate the intact ones to the city. They reasoned that the street clocks "had always been maintained on the public streets, had always proved a public benefit and had ever been demanded by the public."[47] The significance of this disturbingly violent destruction of clocks in Milwaukee lies not in the outlandishness of the episode, but in the jewelers' response: in casting their clocks as public property, rather than advertising, the jewelers harkened to earlier times when they could and did claim temporal authority. As if to shield themselves from accusations of self-promotion rather than selfless public service, three Milwaukee jewelers soon ordered post clocks from Seth Thomas with the words "Standard Time" prominently displayed on the dials.[48]

Now that I have laid out some of the civic contests for authority over the time that simmered and erupted during the public clock era, I turn to the national stage. During the public clock era, the federal government deployed timekeepers in its wide-ranging effort to establish what Lois Craig has called "the federal presence." Few federal clocks were installed in the first part of the nineteenth century, because, as Craig explains, "the federal government was hard pressed to provide physical evidence of its expanding sovereignty."[49] After midcentury, building programs featuring prominent spots for the installation of clocks assumed an important role in the effort to establish federal sovereignty, authority, and stability. Many of these timepieces were E. Howard's popular No. 70 clock; more than 1,200 of them with foot-wide dials and two-and-a-half-feet-high oak cabinets, were delivered to lighthouses, post offices, customs houses, and courthouses.[50]

After the country divided into standard time zones, the volume of federal government orders for public clocks increased considerably. Thirteen percent of its total Gilded Age orders were placed between November 1883 and December 1885.[51] During these years, federal agencies began in earnest to install public clocks throughout the nation. The nation's most politically powerful and economically prosperous regions, the Northeast and Midwest, were the first to receive grand federal clocks. A quick glance at just the Midwest is illustrative. In 1884, fancy and expensive regulators were ordered for Topeka's courthouse and post office and four "United States Clocks" were acquired for St. Louis's customs house. Four clock dials were installed on the tower added to the customs house and post office in Des Moines in 1888. In 1889, a tower clock meant to run hands on three six-foot illuminated dials was destined for the post office in Council Bluffs, Iowa.[52]

As the inventory of federal buildings swelled in the 1890s under the federal buildings program, so did the orders for clocks.[53] Thirty-eight clock orders alone were placed with E. Howard and Seth Thomas in 1898, sending clocks to customs houses spread from Alexandria, Virginia, to Sitka, Alaska; to federal courts in places as remote as Carson City, Nevada; to the US Appraiser's storehouses in New York City and San Francisco; and to the Library of Congress and the new post office building in Washington, DC, now the Old Post Office Pavilion.[54] Only a few federal clocks could be found in the Far West in the 1890s, none of them on the exterior of what often were the only substantial buildings in a town.[55] To amplify the nationalist potency of federal clocks, they were surrounded with flags, framed with sculpted national symbols like eagles, installed with or near representations of historic figures, and connected to large bells that rang the hours.

The trend toward adorning clocks with nationalist imagery reached a fevered pitch around the time of the Spanish-American War. In 1898, President William McKinley's postmaster general, William L. Wilson, publicized his plan for an imposing post office in the capital city. From it the largest flag in the country would unfurl from a hundred-foot-high pole set on a thirty-story tower fitted with four illuminated clock dials and heavy bells.[56] Countless post offices had been erected or leased across the United States, helping to push the country toward conforming to nationwide time standards and schedules. Until the 1890s, post offices rarely "waved the flag"; that is, few prominently displayed patriotic symbols and signs. Clocks could be found inside most post offices, but only a handful were hung outside. The plan for a prominent post office in Washington, DC, crowned with a tower clock and topped with a flag, was part of the larger federal program of erecting public buildings and public devices meant to disseminate the time. It was also in keeping with the post office's own effort to shore up its centrality as a government institution. When Washington's landmark post office with its gargantuan clock tower was completed in 1899, one senator described it as "a cross between a Cathedral and a cotton mill."[57]

Interior of the Old New York Customs House, 1907, New-York Historical Society. A clock with an eagle perched atop it presides over the clerks who oversaw imports into New York City. It authorizes the assessment of duties and taxes, bringing the activities of the busy customs house into focus as government approved.

The close association of clocks with nationalism is highlighted by the addition of four large clock faces set among other national symbols to the US Customs House in Boston, one of the most prominent federal buildings outside the District of Columbia. When appropriations were made for the building in 1835, the winning design included a dome, rotunda, and portico; the building could easily have been a state capitol or part of the national capital's program of architecture. None of the first customs houses had exterior clocks or belfries. Boston was not an exception, despite the nearly one million dollars in appropriations for the building. Instead, its grand view of the Boston harbor, wide staircases, and classical design sufficed. Eighty years later, much of the harbor had filled in, warehouses engulfed the customs house building, and the public square that had once been its frame served as a parking lot. With few other federal buildings in evidence, a grand clock tower would distinguish the customs house from its surroundings. Rising five hundred feet, nearly four times higher than any other city building in Boston, the Peabody and Stearns tower transformed the classical building into a "government skyscraper."[58]

US Customs House Tower, Boston, 1915, courtesy of the Print Department, Boston Public Library.

The dials on the Customs House Tower clock spanned twenty-two feet, covering a floor and a half; the gold-colored glass Arabic numerals were three feet high; the hands, fourteen and eleven feet long. The emblem of the US Customs Service—the eagle—perched beneath each face of the clock. On two sides of the tower, the national bird looks up toward the clock face, while on the other two it looks outward over its domain. The eagles give voice to the national motto "one out of many," while the pyramidal cap for the tower recalls the reverse of the national seal, with its incomplete pyramid captioned "He has favored our undertakings" and "a new order of the ages." The eagles, pyramid, and clock serve as unifying temporal markers, linking the past

("He has favored our undertakings"), present (clock time), future ("one out of many"), and eternity ("a new order for the ages").[59]

More noticeable than the eagles, pyramids, and other signs of the nation that adorned the Boston Customs House clock was the American flag. Draped over and around clock faces, hitched up poles adjacent to or above clock towers, waving alone nearby, flags dressed up the full range of public clocks, not just those of governmental agencies. By grouping national symbols, flags, and clocks, the message was unmistakable: the state's authority for time was inviolable. Few seem to have contested this authority. The abundance of visual evidence from the late nineteenth and early twentieth centuries suggests that wherever a public clock could be found, a flag was nearby. Mechanical time and national pride went together. There were regional variations to be sure, but by the second decade of the twentieth century, the clock and flag formed a composite symbol of the nation.

Time balls also contributed to the effort to extend the federal presence. Naval officers, commercial leaders, and politicians supported the erection, maintenance, and operation of time balls with a passion that did not match the device's limited utility. Despite one period report that described Boston's time ball as being "of use to hundreds of people who regulate[d] their watches by it," US Hydrographic Office evidence from various ports suggests that few relied on time balls to rate their chronometers. In the mid-1880s, a prominent member of the Boston Chamber of Commerce had never heard of the time ball, while city officials preferred to follow the bells "struck all over the city at noon by the Cambridge Observatory." What is more, several ship captains confirmed this lack of reliance, claiming, "They seldom see the ball, and never think of rating their chronometers by it."[60] In the spring of 1885, when branches of the US Hydrographic Office began to offer free rating of chronometers in its Boston offices, years after it had inaugurated time ball services, chronometer merchants expressed outrage at the sabotaging of a chief part of their business. As chronometer maker William Bond and Son complained after the service was in operation a few months: "[It has] become a serious evil and injury to the business." So Bond's storied business, along with those of "Jas. Munroe & Son, Wm. E. Haddock & Co. and thirty merchants & ship owners," sent a protest to the chief of the Bureau of Navigation.[61]

Since the first drop of the Western Union time ball in 1877, William F. Gardner, the leading instrument maker for the US Naval Observatory, had been working to perfect the mechanism for releasing the ball when the telegraphic time signal was received, which was acknowledged as "the essential part of the apparatus."[62] The ball, with a diameter between two and four feet, was constructed out of a framework of wood, steel, or iron and covered with canvas, usually either black or red, and weighted sufficiently so that it would drop instantly fifteen to twenty-five feet.[63] Shortly before noon, a technician would manually raise the ball up a staff secured to the top of a building or tower. A few seconds before noon, he would apply the releasing apparatus (which utilized a magnet and electrical circuits), and if all went well, an automatic signal would then release the ball.[64] With such a setup, it is easy to understand how things could go awry: wind or ice could prevent the ball from being hoisted; trouble with the wires could interfere with the signal; the releasing apparatus could be triggered accidentally. Reports from officers who managed these contraptions further highlight their unreliability: frequently the balls did not even drop, owing to faulty signals, poor weather, and operator error.[65] Even after Gardner's releasing apparatus was perfected, more than one officer in charge of time balls made it his "custom to drop the ball by hand," surely not a reliable way to transmit accurate time.[66] When the ball was released at the wrong time, most officers turned to the braking system to "arrest the ball in its descent," and later to the local press to notify the public that the ball had been dropped in error.[67]

Time balls denoted aspiration in scientific, financial, and cultural arenas, particularly a willingness to discard local time in favor of the standard of time zones. By 1884, the US Navy had secured its long-held right to erect and operate time balls, overcoming a legislative challenge posed by the US Army Signal Service, which proposed to erect time balls at fifty-five of its signal service stations (including ones in Bismarck, Dakota Territory; Mineral Wells, Texas; and Santa Fe, New Mexico Territory) and to cooperate in the maintenance of "public Standard Time Balls" elsewhere. The navy presented its own plan, which involved placing time balls on customs houses at ports of entry and in other cities of more than 15,000 residents.[68] Neither of these plans received sufficient political support, most likely since appropriations would have had to amount to more than $25,000, about half a million dollars in today's dollars, during a time when the federal budget covered little more than lighthouses and military pensions.[69] But the navy did secure the right to erect time balls where it could without additional appropriations; the US Naval Observatory and the US Hydrographic Office oversaw the construction, maintenance, and operation of time balls.

These intergovernmental contests for authority over time balls coincided with the short-lived relocation of Washington's time ball from the US Naval Observatory to the State, War, and Navy Building. This was not happenstance, but a deliberate choice meant to communicate the nation's veneration of accurate time and its growing aspirations as a world power.[70] Around the

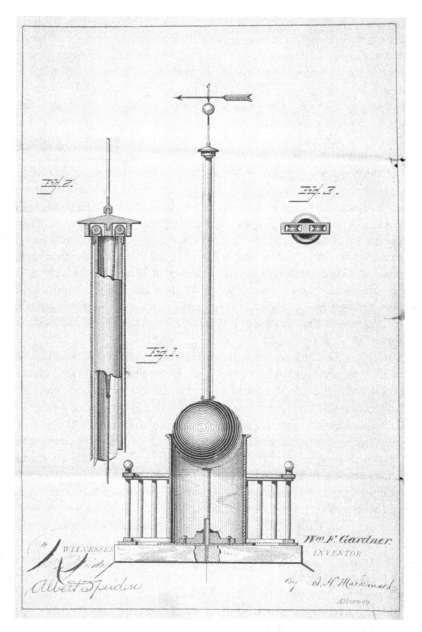

Diagram from William F. Gardner's application for a time ball patent, 1884, from Boston file, in "Time Ball Reports," Records of the United States Hydrographic Office, United States Navy, National Archives Records Administration.

same time, a proposal to erect a time ball on the Washington Monument, which was still under construction, was presented. An unsigned letter to the US Naval Observatory made a case for "the propriety of placing a time signal on top of the Washington Monument when it shall be completed." What better signal than a time ball dropped according to observatory time? The letter explained why: "[The] signal would be useful as furnishing a reliable means of setting time pieces over an extended area." In particular, it was the "use of the monument" to disseminate time that the letter's author suggested would be "in harmony with its time enduring grandeur."[71]

The desire of some places to adorn their newly developing skylines with symbols of accurate time, national prestige, and imperial ambitions coincided with the expanded means of the navy as well as other federal agencies. Under a variety of federal auspices, for example, Philadelphia, Baltimore, Portland (Oregon), Savannah, Woods Hole, and Hampton Roads (Virginia) acquired time balls. The Fish Commission sponsored the Woods Hole time ball; the Torpedo Station sponsored the ball in Newport, Rhode Island; the US Hydrographic Office sponsored the time balls in Philadelphia, Baltimore, New Orleans, and on San Francisco's Telegraph Hill; and the US Naval Observatory sponsored the time balls in Washington, DC, Hampton Roads, and Savannah.[72]

How New Orleans secured permanent time ball services is instructive about the partnerships that fostered clock time. The US Naval Observatory sent a time signal every day at noon that released a temporary time ball on top of "the central tower of the Main Building" at the 1884–85 Cotton States Exposition. It also regulated "a great many clocks" at the fair and was responsible for "a Torpedo exploded" in the exposition grounds' lake.[73] As the dismantling of the exhibition neared, several proposals for erecting a permanent time ball appeared in local papers, with justifications focusing on "the value of precisely correct standard time."[74] In June 1885, a time ball was hoisted atop a twenty-foot pole on a fifteen-foot tower rising from the roof of the New Orleans Cotton Exchange, placing it 140 feet above sea level. The ball could "be seen from the balconies, upper windows, or roofs of nearly all houses in the city and from nearly all parts of the river front, either from the decks or riggings of ships."[75] Beneath the ball was the Cotton Exchange's enormous bell, which was struck twelve times each day at noon. In this way, the federal government provided both visual and aural signals conveying accurate time to postbellum New Orleans.

This New Orleans tribute to modern times underscores the monumental, rather than the utilitarian, aspect of time balls. Without any notice in the Crescent City's newspapers, "a very large crowd gathered on the streets to witness" the first noon drop of the time ball in late June 1885. The next day

New Orleans Cotton Exchange Building, ca. 1903, LC-D4-16323, Prints and Photographs Division, Library of Congress. Next to the thick smokestack rising from the roof, there is a pole for the time ball. Clock merchant Maurice Scooler likely installed the large clock dial.

"an immense crowd on the streets" again waited "to see the Ball drop." But trouble with the telegraphic signal prevented the spectacle from unfolding.[76] For the next quarter century, a naval officer, or someone he dispatched, oversaw the Cotton Exchange's time ball. Sometimes it would drop at noon; other times it would not. The monthly reports documenting the erratic performance of the time ball that were filed for several years throw into doubt the New Orleans Maritime Association's assurances, stated in a letter to the chief of the Hydrographic Office, that "all the principal jewelers, chronometer makers etc. take advantage of [the time ball's] indisputable accuracy." Nor is it possible to take as a fact the association's additional endorsement that "many strangers and residents compare their timepieces daily upon the dropping ball."[77] It is unlikely that more than a few people set their timepieces by the erratic time ball; that it imputed to the once thriving port of New Orleans a semblance of efficiency and punctuality seems more likely.

Shortly after the turn of the century, the US Hydrographic Office received many requests for time balls. During the first decade of the twentieth century, the navy initiated time ball service in Baltimore, Chicago, Cleveland, Philadelphia, Sault Saint Marie (Michigan), Buffalo, Galveston, Duluth, Norfolk (Virginia), Key West, and at the Cavite Naval Station in the Philippines.[78] That urban Americans watched for the drop of the ball is documented. The hydrographic officer in Portland, Oregon, noted in 1909 that he saw many people "standing in the park, on the corners, etc. etc. at noon waiting for the ball to move." He surmised that this showed "the close watch people keep on the ball."[79] Time balls may have helped people regulate their watches, but the only navigational direction they provided was to avid participants in the consumer culture, for with the exception of the naval base in the Philippines, they were sited on the roof of each community's most prominent commercial building.

This turn toward partnership with commercial agencies continued. In 1909, the navy sponsored the construction of a time ball on the roof of San Francisco's Fairmont Hotel. In 1910, it installed a time ball on the floor of the Philadelphia Bourse. In 1911, the navy agreed to fund the erection of a new ball on top of the Maryland Casualty Company Tower, which when completed in 1912 would be Baltimore's tallest building, and one of the taller buildings in the nation.[80] Even though a third of the navy's time balls were erected after wireless transmission of time signals was possible (1903), it only fulfilled a small number of the requests for time balls. Civic and business leaders across the country unsuccessfully petitioned for time balls: their ranks included a large wholesale and retail business in Rochester, New York; the Knoxville Tennessee Bank and Trust; the Chamber of Commerce for Pensacola, Florida; the Honolulu Photo Supply Service; the Board of Trade of Jacksonville, Florida; a Salt Lake City investment company; Tampa's Board of Public Works; a congressman from Alexandria, Virginia; the mayor of San Diego; and a newspaper in Lynn, Massachusetts.[81]

Despite this burst of construction and solicitation of even more projects, the short-lived vogue for time balls was coming to an end. Late in 1909, a tropical storm carried away the New Orleans ball.[82] A few months later, in 1910, the mast and ball on the Cotton Exchange were dismantled. For several years letters were circulated, meetings held, and surveys commissioned in order to determine where the new time ball should be erected. New tall buildings overshadowed the Cotton Exchange, and the navy wanted to ensure navigational use of the time ball by moving it to a site visible from all docks, which was almost impossible given the city's location on a crescent-shaped bend in the Mississippi River. But the time ball had never served as a navigational device, and with the introduction of wireless transmission of time signals, the effort to impute navigational value was hopeless. Eventually it was decided that it did not need replacing.[83]

Around the same time, speculation percolated about where to move New York City's time ball, since the one on the Western Union building was no longer visible. The hydrographic officer stationed in New York City observed: "A visible noon time signal is thought to be desirable solely for the sentimental reason that this great port should not be without one." He explained further: "Observatory Time clocks are to be found in nearly every office in this City; . . . wireless noon signals can be used by almost every steamer; . . . nautical firms, not only take charge of and rate chronometers, but also when requested send Agents to ships to correct chronometers." Moreover, he added, "the most conspicuous and best known location in the City," the Metropolitan Life Tower, had a "thoroughly admirable night signal."[84] In the end, a new ball would go up on the Titanic Memorial Lighthouse Tower built near the tiny Seaman's Church Institute Building on South Street in 1913. The Western Union time ball was discontinued the following year, the apparatus eventually finding its way onto the Titanic Memorial Lighthouse Tower.[85]

During the First World War, the Hydrographic Office dismantled most of its time balls. In 1920, it turned down a request from the governor of the US Virgin Islands for a time ball, telling him that radio service was superior.[86] By 1922, only a few time balls were in service, and in 1936, the US Naval Observatory's own time ball was "decommissioned." The time ball was a tribute to instantaneousness and accuracy, an indication of federal and civic eagerness to embrace technological fixes. Soon enough, other monuments to technological prowess—skyscrapers, giant clocks, electric signs, and wireless transmission of information—had displaced it.

Through many different tactics, mostly uncoordinated and unplanned, the federal government staked a lasting claim to ownership of time in and across the United States. Beyond the fact that it erected and maintained public clocks and time balls, encouraged the nationalist decoration of clocks, and funded the time-reckoning and time-keeping operations of the US Naval Observatory, several other developments further asserted the federal government's proprietorship over the time. First, Congress officially adopted 75th-meridian time for Washington, DC, in 1884. Second, that same year, the US government hosted an international meeting (the International Meridian Conference) at which discussion centered on the adoption of a universal time system with Greenwich as the meridian. Third, some years later, in 1918, Congress declared standard time national law (while at the same time mandating daylight saving time as an energy-conservation measure).[87] Federal, state, and local government shared jurisdiction over the time. However, the ultimate authority for it rested in government agencies headquartered in Washington. The US Naval Observatory and later the Bureau of Standards determined the correct time. Various branches of the military, the Interstate Commerce Commission, and, after 1966, the Department of Transportation disseminated the time, oversaw time zones, and implemented daylight saving time.

Time balls and tower clocks were among the first material representations of the federal government outside the District of Columbia. They appeared in cities during the same decades that soldiers' homes were spreading across the nation, monuments to American presidents were being dedicated, federal courthouses were being built in each federal jurisdiction, and a long-term program of lighthouse construction (initiated by a 1789 act of Congress) was accelerating. The number of lighthouses alone doubled between 1880 and 1910, from around six hundred to twelve hundred.[88] Time balls and tower clocks, like coastal beacons, connected local communities to Washington, DC, the seat of national power, and to clock time, a source of national power. During the same decades as federal investment in national timekeeping deepened, sponsors of monumental clocks, monster clocks, and memorial clocks further inscribed clocks with national significance and potency.

CHAPTER 6

Monuments and Monstrosities

The Apex of the Public Clock Era

Won't the Public Buildings Commissioners kindly get together and agree on a standard of time, so that all four faces of the tower clock may be made to tell the same story? This thing of giving all times to all men may be in conformity with the political methods at the Hall, but it's disquieting to a citizen to find that it is quarter of five on the north side of the Hall and on the south side seven minutes to six, while the west view shows ten minutes to eight, as against three minutes to four on the east.

— "What Time Is It?" *North American* (1899)

n 1899, Philadelphia's city hall had been under construction for decades: the situation was one of the notable examples of the graft and corruption that accompanied the growth of American cities. Its clock, one of a handful of what were called "monster clocks" because of their enormous size, had been set going at the first instant of the year, but the four twenty-five-foot dials rarely showed the same time.[1] By the end of the nineteenth century, Philadelphia, like other cities in the United States, was saturated with clocks and bells. The town hardly needed the gargantuan clock, but no matter, since practicality was low on the list of city priorities. Railroaders, merchants, jewelers, factory managers, city officials, government bureaucrats, bankers, and brokers, among others in Philadelphia, had already installed scores of clocks on buildings within a few miles of city center. Among the orders they placed with Seth Thomas and E. Howard were requests for clockworks to operate an astronomical regulator attached by electrical circuit to more than two dozen wall clocks for the US Courthouse and Post Office, façade and tower clocks for the insurance and banking office towers going up on Chestnut Street, a wooden double-dial clock to hang in front of the Estley Organ Store, a four-faced post clock for the Reading Railroad terminal, the tower clock of the Craft and Allen Candy Factory, and timekeepers for a new market house and a new police station.[2] Confidence ran high that these clocks would convey legitimacy and authority, rather than become sources of embarrassment with stopped or errant hands.

Just when the challenge of setting clocks to a time standard had been met with time circuits and time services, a new impediment to synchronicity developed. In the effort to heighten visibility and audibility, the sponsors of public timekeepers set enormous dials with massive hands high upon towers. It was nearly impossible for the dial works of these monster clocks to function reliably. Imagine a technological device aiming for precision and accuracy that depended on "a little puff of air" traveling up seven floors "through a leaden tube having an inner diameter of no more than the sixteenth of an inch." This is what was required to move forward "the ponderous minute hands" of the Philadelphia City Hall clock "about seven and one-half inches" every thirty seconds.[3] The four dials on the clock showed different times: there was not enough motive power behind the puffs of air sent from the clockworks more than three hundred feet below to move all hands in unison. By the 1920s, electric motors and other devices would solve this problem, but by then other more reliable methods of disseminating accurate time, like radio and telephone time services, had been introduced. At the turn of the century, monster clocks seemed a good way to disseminate the time to the nation's largest cities: the populations of Philadelphia, New York, and Chicago each surpassed one million. How else could a single, reliable time signal reach so many people?

As the first seat of the US government, not to mention the place where the founding documents were conceived, Philadelphia was where the nation's time began. Its long history as a center of scientific inquiry and precision instrument manufacture made the city an early site of American clock and watch manufacture. By the 1880s, Philadelphia was home to several watch-case factories, a few high-end chronometer makers, and an active retail and wholesale trade in pocket watches and all manners of clocks for homes and offices. It is no wonder that in 1899 the faulty timekeeping of the city's most grandiose clock drew controversy. Exacerbating this trouble were plans to take down Independence Hall's tower clock dials and restore the colonial-era tall-case clocks, though without clockworks. The vogue for "monumental clocks," in which painted scenes, automatons, and clock dials enacted primal scenes from American history, enhanced interest in historic timekeepers like those of the Old State House. With the exception of the Liberty Bell, relics these clocks could not become. They had to work—their hands had to move continually, their hour's bells ring faithfully—to convey the full extent of the nation's destiny.

This chapter explores the powerful claims about the meaning and importance of clocks and clock time circulating at the apex of the public clock era. It considers how monster and monumental clocks connected ordinary Americans, most of whom owned one sort of mechanical timepiece or another, to a scheme in which clock time took on historic and cosmic meanings.

By the 1893 world's fair in Chicago, a national system of standard time, an extensive grid of public clocks, and a mass market for watches were in place. The complexity and extent of the fair's exhibitions related to timekeeping were stunning. A time ball adorned the US Government Building, as one had at earlier expositions held in the 1880s.[4] In a small building next door, the US Naval Observatory exhibited an astronomical clock connected to an electrical time circuit. The display showed how the observatory sent accurate time to seventy thousand clocks and nine time balls located throughout the United States.[5] In the Manufactures Building, the Self-Winding Clock Company's timekeeping mechanism ran the hands on seven-foot dials hanging from a 140-foot-high clock tower. It operated an apparatus responsible for striking the quarters and hours on bells weighing altogether fourteen thousand pounds. The Self-Winding Clock Company's master clock was also synchronized to adjust two hundred self-winding clocks found throughout the fairgrounds. In still another virtuosic display of timekeeping, the fairground's Central Railroad Depot featured a frieze of twenty-four clock faces, each five-feet in diameter, showing the time of the world's leading cities.[6]

Historic watches and clocks were included in various exhibits. The watch Miles Standish carried on the Mayflower was displayed in the Mount Vernon Building. The Virginia Building featured "an old Washington family clock." The Pennsylvania Building, where the Liberty Bell was on view, replicated Independence Hall's clock tower. The US Naval Observatory flourished "chronometers of historical interest and specimens of the best of American manufacture." The Daughters of the American Revolution presented to the fair's commissioners the Columbian Liberty Bell, which weighed thirteen thousand pounds, a thousand pounds for each of the original thirteen states.[7] The message was clear, Americans were a timekeeping people.

Not only that, they manufactured their own timepieces, and much of the world's as well. The industrial production of precision instruments in the United States excelled as had no other national industry before it. In a "prominent location on [the fair's] Columbian Avenue," the Waltham Watch Company's exhibition included "automatic machines for making watches," replicas of its factory buildings, historic watches, as well as its complete line of cased and uncased pocket watches. It was an awe-inspiring display: one diarist described seeing "hundreds of watches in one case all gold." The Ingersoll Watch Company sold as many as 85,000 watches for $1.50 each at the fair. Other watch companies who exhibited at the fair included the Waterbury Watch Company, the Keystone Watch Case Company, the Ray Watch Case Company, and numerous French, Swiss, and German watch manufacturers.[8] In all, the various technological and symbolic components of modern time discipline were on display and for sale at the world's fair.

Several monumental novelty clocks stand out from this array of impressive exhibits related to timekeeping at the Chicago fair. Tiffany's display included an eight-foot-high astronomical clock with twenty-five silver and enameled dials. Highlighting only two years in world history—the year the Julian calendar was introduced and the year of American independence—its precious dials showed local time, Washington time, Greenwich time, and the time of thirty-one world cities. The Waterbury Watch Company's exhibit featured the twenty-foot-high "Century Clock" with automatic figures. Paying homage to the United States was an extraordinary timepiece constructed by the Goldsmith's and Silversmith's Company of London. Eight panels on the eight-foot-high gilded case of the "Exposition Clock" represented seven sports and the Brooklyn Bridge, above which hung portraits of Washington, Lincoln, Grant, Franklin, Jackson, Harrison, Cleveland, and Queen Victoria. An American eagle presided above each of the four dials, which showed "American, English, French and Spanish time." When the clock struck the quarter hour, twelve figures playing sports revolved; when it reached the hour mark, American and English anthems played.[9] Tiffany's astronomical clock, Waterbury's Century Clock, and the Goldsmith's and Silversmith's Exposition Clock each presented "American time" as a mixture of historic iconography and mechanical time. The eye-catching large clocks, cases of watches, and shelves of decorative timekeepers, along with the synchronized time systems in place, were prominent in the grand arena that was the 1893 world's fair. These exhibits of timekeeping prowess bolstered the enactment of imperial fantasies and racial hierarchies characteristic of both this and other international expositions.[10]

Traveling exhibits of monumental novelty clocks further capitalized on the association of historic time and clock time, while also supporting notions that clocks were civilizing and nationalizing mechanisms. The first known American monumental clock, the Engle clock, was completed in 1878 after twenty years of labor and then sold to an itinerant exhibitor. Over the next three-quarters of a century, a number of enterprising showmen charged fifteen to twenty-five cents for a viewing of what they considered "the Eighth Wonder of the World." Modeled after Apostolic clocks on display in Europe, particularly the Strasbourg clock completed in 1352, Stephen Engle's eleven-foot-high clock included the figures of Jesus Christ, the twelve apostles, Satan, Father Time, Orpheus, and a regiment of Continental soldiers marching to battle.[11] In the same category was Detroit clockmaker Felix Meyer's "remarkable clock" completed in the late 1870s. It was housed in a black walnut cabinet engraved with "designs appropriate and symbolic of our Republic," stood eighteen feet high, and weighed five thousand pounds. One of the many notices for the clock described George Washington sitting in a marble dome above the clock's body: When Father Time strikes the hour, then "Washington slowly rises from his chair to his feet." As he presents the Declaration of Independence, all the presidents file by and salute him. When the last president (Hayes) disappears, "Washington retires into his chair and all is quiet, save the measured tick of the huge pendulum and the ringing of the quarter hours."[12]

By the end of the century, there were several dozen such clocks on permanent and traveling display.[13] William Robert Smallwood built a sixteen-dial world clock between 1884 and 1891 that was displayed at the Columbian Exposition in 1893. The clock's middle dial showed standard time, while the dials arrayed around it showed Greenwich time, along with the time of such places as Rio de Janeiro, Mt. Ararat, Denver, San Francisco, the Sandwich Islands (Hawaii), Calcutta, and Peking.[14] A Canadian-born engineer, Myles Hughes, built a seven-foot-high clock between 1881 and 1916 in Buffalo that showed standard and natural time on its dials, and whose figures included Jesus and the apostles.[15] In the 1890s, the thirteen-foot-high "Great Historical Clock of America," now part of the Smithsonian Institution's collections, toured the United States, Australia, and New Zealand with a blackface troupe known as Christy's Minstrels. Flanking the wooden case displaying two hundred historic American figures, a clock dial, and a zodiac, were replicas of the Statue of Liberty and the soldier's monument at Gettysburg. Like the Engle clock, at every quarter hour a figure of each American president, ending with

The Engle clock, ca. 1878–88, photograph in the Library of the National Association of Watch and Clock Collectors, Columbia, Pennsylvania. The Engle clock toured the Atlantic seaboard for nearly seventy years, before disappearing for three decades in 1951. It is now on display at the NAWCC Museum.

Benjamin Harrison, made an appearance. An eagle perched atop the case, while another one with a flag was painted near its base.[16]

The interest in monumental clocks continued into the twentieth century. Chicago clockmaker Frank Bohacek gained notice in 1908 for his mammoth celestial clock. The brothers Frank and Joseph Bily, lifelong residents of Spillville, Iowa, constructed nineteen monumental clocks between 1913 and 1948, including the "American Pioneer History Clock" with fifty-seven panels showing events from American history. Fascination with clocks also manifested itself with the acclaim for miniature reproduction clocks made by

Andrew William Marlow (1903–84) in the late 1930s and early 1940s. Marlow, a furniture maker in York, Pennsylvania, made fourteen different kinds of replica clocks, from the No. 1 Simon Willard to the Half Moon Grandfather, whose movements were made by the New Haven Clock Company. He might have made as many as several thousand, some of which were sold at Macy's and B. Altman's department stores. As Carlene Stephens and Michael O'Malley explain, monumental novelty clocks "equated time with American history and progress."[17]

So too did the thousands of public timekeepers that saturated American public spaces with aural and visual time cues during the decades around the turn of the twentieth century. A line of caricatured figures—including an African with a spear and a Native American in a headdress—standing in awe

J. S. Johnston, *New York Herald Building*, 1895, Photo Archive of the Museum of the City of New York. Newspaper buildings' extravagant timekeepers, such as the one installed on the Herald Building in 1895, characterized the era of monumental public timekeeping. The goddess of wisdom, Minerva, presided over the building's bell and its two five-foot dials, one showing the time, the other a compass. When the bell rang on the hour, it appeared to be struck by the two blacksmith figures. An outsized variation on popular novelty clocks, the Herald clock drew attention in a city crowded with timekeepers. Its sculpture "Minerva and the Bell Ringers" remained in Herald Square after the Stanford White building was demolished in 1921.

in front of a post office topped with a prominent clock tower found in an 1888 design for the US Postal Service reveals the extent to which federal buildings and timepieces were meant to assert authority and reinforce racial hierarchies.[18] As Anglo-Americans built a hierarchy based on national origins, race, and nativity, they reified and invested in symbols that reinforced their sense of superiority. Not just clocks and watches, but apparent harmony between them and society, was one such symbolic system. Anglo-Americans came to believe that other peoples lacked a sense of time, or if they had one, that it was a "primitive" time sense. They assumed that they lived in time; in their view, everyone else in the world except Europeans, lived out of time. Without clock time various states of savagery were certain.[19] Living in time was an indicator of civilization, so much so that photographers and other ethnographers of "primitive" peoples sometimes removed evidence of watch and clock ownership from their records.[20] The hegemony of mechanical time

Edward S. Curtis, *In a Piegan Lodge*, 1910, untouched negative, LC-USZ62-61749, Prints and Photographs Division, Library of Congress. The ethnographer Curtis removed the prominent alarm clock from this photograph when he published it in 1911 in the sixth volume of his monumental series, *The North American Indian*. Had he left the alarm clock in the photograph, it would have contradicted prevailing ideas about Native Americans as "primitive" people.

was so overwhelming by the end of the nineteenth century that scant regard was paid to how time might have been kept without clocks and watches.

"Clocks are what is troubling the official mind these days," opened a June 1898 Philadelphia newspaper editorial. The city was absorbed by two major clock projects. One was the "monster timepiece" destined for the tower of the grandiose Public Buildings (the official name for the city hall); the other concerned the tower clock at Independence Hall, where working from engravings dated 1800 and 1815, the architect T. Mellon Rogers planned to replace the dials with decorative windows of his own design and install tall-case clocks on the building's exterior.[21] To justify the dials' removal, the city's director of public buildings flourished historic engravings depicting the state house of the Revolutionary period, called upon the memory of a "foremost citizen," quoted city council records, and presented archaeological evidence—all demonstrating that the tower clock was a modern addition to the historic building. Despite such armature, the very suggestion that the tower's clock dials might be removed "aroused opposition."[22] As one newspaper writer observed, "the transference [of clock dials to the tower in 1829] was a public convenience, and a public convenience it has been ever since." The "clock in the tower" had been hailed for decades as "an institution in Philadelphia."[23]

After the controversy had brewed for a month, the director of public buildings again defended removing the clock. He explained that it was necessary in order to restore Independence Hall to its 1776 condition. One of Philadelphia's many newspapers supported the plan, commenting that, "it is a work of restoration, not modification." Denigrating the utilitarian purposes of the tower clock and bells bolstered the case for their removal: it was claimed that there was no longer any need to strike fire alarms on its bells, "nor would anyone today think of setting his clock or watch from a State House clock."[24] With the controversy about the clocks so heated, no change was made prior to the rededication of Independence Hall in October 1898. Engravings depicting the celebration show the steeple clock quite prominently, as if the engravers and printers themselves were expressing an opinion on the matter.[25]

Despite its central place in the two most important temporal regimes (historic time and mechanical time) of turn-of-the-century Philadelphia, the Independence Hall clock was acknowledged as "the second most important time-piece in Philadelphia."[26] Its importance dimmed in comparison to the city hall's monster clock. A variety of efforts over several years led to the moment when the enormous tower clock began its work. After months of investigation, which took a committee of commissioners of public buildings

William Birch and Son, *Back of the State House*, Philadelphia, 1799, engraving, courtesy of the Library Company of Philadelphia. Elegantly dressed women and men stand below the imposing clock on the exterior wall of the state house; just paces away are four Native American men, set apart not only by their buckskin garments but also by the red feathers instead of hats on their heads.

to Milwaukee, Minneapolis, Newark, and Washington, DC, the commissioners selected a novel and largely untested pneumatic system. Compressed air would move the clock's massive minute hands (each weighing 225 pounds) and hour hands (a mere 175 pounds each). The four clock dials would be connected to a master clock with "all the latest and most approved improvements to insure accurate time," including being on a telegraphic circuit to the US Naval Observatory, which, as we have seen, was considered the nation's most reliable source of time.[27] While of an unusual scale, Philadelphia's city hall clock belonged to the Gilded Age culture of public timekeeping in which clocks on civic buildings proliferated.

Shortly after the rededication of Independence Hall in the fall of 1898, commemorative friezes on the new city hall tower surrounding four twenty-five-foot clock dials were completed. Rising 361 feet from the building's top floor, the tower's height soars above the bell tower of St. Mark's in Venice and other landmark buildings in Western Europe. Bronze eagles perch above the clock dials. Crowning the entire ensemble is a statue of William Penn so large that according to a period publication its mouth "would easily take in a whole turkey in one bite."[28] Allegorical groups "representing four epochs in the early history of Pennsylvania" sit next to the pediments framing the clock dials. The eagles are particularly noteworthy, since throughout the nineteenth century they served as a national motif on timekeepers of all sorts.[29] Plans were made to illuminate the dials of the clock, so that the time could seep across the city through night hours. Furthermore, the entire tower itself was meant to serve as a time signal, with its lights extinguished momentarily to indicate nine o'clock each night. Even at a distance, then, watches and clocks could be reset. This custom "united the sprawling grid of Philadelphia," as art historian Michael Lewis observes, in the same way "cathedral bells once united medieval towns."[30]

Throughout the fall of 1898, Philadelphians marked the progress of their large clock's installation on the city hall, attributing timelessness to it even before it started ticking. Although the clock was hung and ready in early December, the commissioners decided to wait until New Year's Eve to set it going. That night a party was held for the building commissioners, the mayor, and other officials, who then witnessed the starting of the clock.[31] They were but a small portion of the people interested in the new clock; thousands crowded the streets around the city hall. "On the stroke of 12 last," it was reported, "pealing bells, the shrill notes of horns, and shouts and cheers of thousands" welcomed "the last year of the nineteenth century." Emphasized here are the aural sounds of time—bells and horns—heralding not only the New Year, but also the clock's debut. But its visibility was also of note. It was reported that the assembled crowds "in the City Hall square and the surrounding streets and avenues" saw "the great dial of the colossal new clock in the City Hall burst into radiant light," and marveled as "the gigantic hands began their unceasing circuit."[32]

By the turn of the century, it was a common practice to look to clocks when a New Year was imminent. Through the 1870s, bells had greeted the New Year, not exactly at the instant of midnight, but approximately around then. And who knew, or cared, if it was the precise moment anyway? But by the 1890s, with the emphasis on precision and accuracy, a clock was an essential accoutrement for counting down the hours, minutes, and seconds until the New Year. It was not just public clocks that were looked to for the auspicious moment, but also those found in homes. Many personal accounts describe listening for the chimes of a mantel or tall-case clock before welcoming the New Year. By the twentieth century, public clocks had gained potent iconic powers when it came to ringing in the new year: whether with commemorative images of small children stepping out of clocks, or toasts made

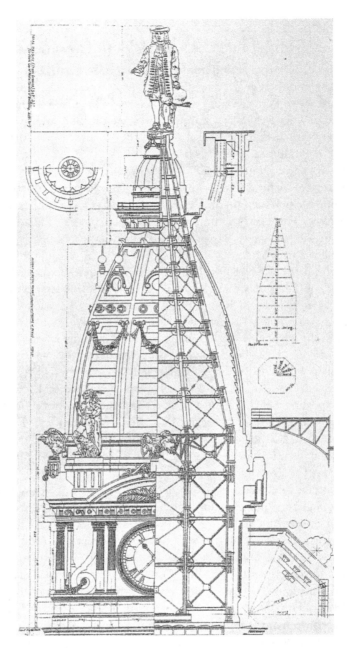

Clock tower section, Philadelphia Public Buildings, 1889, frontispiece for *Contract for Metal Work for Completion of Clock Tower, Philadelphia* (printed for the Commissioners for the Erection of Public Buildings by the Press of Henry B. Ashmead, 1889). This sectional drawing shows the thirty-six-foot-high statue of William Penn, the twenty-four-foot-high statue of an Indian, and the tower's clock face. The tower soared nearly 548 feet above the pavement. Many features of the clock are noteworthy, including the absence of numerals, an innovation at the turn of the century, as well as its $27,000 price tag. It was widely characterized as "a monster timepiece" and drew as much criticism as did the buildings.

Dial work for tower clock, Philadelphia Public Buildings, 1901, in Warren S. Johnson, "The Philadelphia City Hall Clock," *Journal of the Franklin Institute* 151 (February 1901): 100. When installed, this was the largest clock dial in the world, at twenty-five feet in diameter. As clock dials grew larger and larger, pictorial representations frequently depicted men on their dials. This motif lasted through the 1960s and beyond in American popular media.

when the hands hit twelve, or the dropping of a time ball at midnight, clocks and new beginnings were intimately associated.[33] What is more, as arrival of the New Year increasingly became a visual phenomenon, clocks showing midnight became far more important than bells tolling the hour. In many cases, city crowds simply could not hear the bells that tolled when the New Year arrived: their own members were far too noisy with tin horns and other noisemakers.[34]

Harry M. Rhoads, *Child Stepping out of Clock, Charles C. Gates' House, Colorado*, ca. 1916, Denver Public Library, Western History Collection. Depictions of small children stepping out of clocks conveyed the sense of new beginnings that came with the New Year. Motifs such as this one were commonly found on New Year's greeting cards. When not set going on the Fourth of July, public clocks often made their debut at midnight on New Year's Eve. The second child born to the rubber and tire magnate Charles Gates, Hazel, is pictured here stepping out of a large tall-case clock on the stairway of a mountain chateau.

Anticipation of the "unceasing circuit" of the hands on the newest clock in Philadelphia in 1899 stimulated grandiose dreams befitting the largest time-piece in the world. A celestial presence was attributed to it: one writer commented that "the great timepiece was shown like a star." Another asserted that it "was unlikely that the clock [would] thereafter cease its labors for a moment during the lifetime of any person now living." Like the North Star, this clock was expected to "regulate the comings and goings of countless succeeding generations of Philadelphians." It would "serve as the standard and regulator of clocks and watches of all posterity."[35] The clock itself was multivalent: in its statuary, it invoked both Pennsylvania's and the nation's origins; in the anticipation of its perfect operation, it was like the celestial clocks of one of the early keepers of the Old State House clock, David Rittenhouse; and in the emphasis on its visibility and accuracy, it spoke to desires for efficiency, accuracy, and synchronicity.

All of these hopes, however, were dashed within an hour of the clock's inauguration. The hands on the north dial stalled, probably because of the wind, sleet, and hail brought by a winter storm. The following day it was reported, "Broad Street travelers saw the monster hands flinging themselves wildly over the face [of the north dial]." Crowds of people gathered below, "gazing bewildered up at the tower." Reportedly one man on the street was said to have remarked, "Time's flying," while another commented, "It will be tomorrow before this afternoon."[36] All quips aside, the clock did not work. By May the clock's problems were such that the city commissioners convened a meeting to discuss its "pure cussedness."[37] In its inability to show the correct time, in the wild movement of the hands, the clock itself highlighted the artificiality of mechanical time as well as the unreliability of mechanical timekeepers. Its four faces could not consistently keep the time. It could not possibly "regulate the comings and goings of countless succeeding generations." Such an irregular clock had no hope of serving "as the standard and regulator of clocks and watches of all posterity."[38]

Other cities also installed massive clocks around the turn of the century. San Francisco's Union Depot and Ferry House, completed in 1901, had "one of the largest pieces of time-keeping machinery ever constructed." The massive iron dials on the outside of the tower walls supplied time day and night. The day dials were a bit more than twenty-three-feet in diameter, and weighed twenty-five hundred pounds; the night dials were made of a heavy metal framework that was "glazed with thick ground plate glass, arranged with metal figures placed on the glass so that they [were] only visible at night when lighted from inside." The clockworks ran a secondary clock system that operated twenty-one interior clocks.[39] After San Francisco's 1906 earthquake, photographs and engravings of the ferry clock, with its hands stopped

Destroyed clock in front of Baldwin Jewelry Company, San Francisco, 1906, Shaw and Shaw photograph, FN-34712/CHS2012.848.tif, courtesy of the California Historical Society.

at 5:16, proliferated. They joined a visual account in which all vestiges of civilization, including street clocks, were shown beyond repair.[40]

The stopped ferry clock came to symbolize the ravaged city of San Francisco. If a less disturbing image than the craters, mountains of rubble, ash heaps, and thousands of dead bodies, the clock nonetheless conveyed time gone awry. Trying to imagine a new day, one journalist several months later reflected: "To be sure, we look today upon the ruins of our business district and we grieve over the loss of our landmarks. But our harbor is still here, and our docks and our ships. Who shall say that when we note the hour and minute in the ferry tower again it will not be the beginning of a new era in San Francisco's commercial progress?"[41] When the clock was set going again on December 31, 1906—New Year's Eve—the city celebrated. Twenty sailors in a waterfront saloon raised a toast to the ferry clock, with one announcing upon spying the illuminated dials: "There she's beaming. Here's to her." With "twenty glasses raised and drained," the sailors hoped "may she never go out again."[42]

By the first decade of the twentieth century, large clocks like the ones belonging to the Milwaukee City Hall, the Philadelphia Public Buildings, or San Francisco's Ferry House were part of the mature landscape of state activism and corporate promotion. It was the era during which skyscrapers extended the vertical reach of cities.[43] In 1908, the Colgate Company, which made soap, perfumes, and toothpaste, installed atop its Jersey City factory

Bent steel flagpole and stopped clock on San Francisco Ferry House, 1906, FN-23754/ CHS2012.847.tif, courtesy of the California Historical Society. A contrast might be drawn between the purposeful businessmen hurrying beneath the stopped clock in this photograph and the working women primly sitting on the largest clock hand in the world in the 1909 photograph reproduced on page 161.

Colgate Company clock, 1909, Division of Technology, National Museum of American History, Smithsonian Institution. Seth Thomas made this thirty-eight-foot diameter clock face for the Colgate Company factory and headquarters in Jersey City, across from lower Manhattan. When the company replaced it with a clock whose diameter was fifty feet in 1924, this one became the second largest clock face in the world. It was reinstalled at the Colgate Company's factory in Indiana.

and headquarters a clock dial that measured thirty-eight feet in diameter by day, forty feet by night (the lights that were the hour marks were set outside the daytime dial). It was part of an enormous illuminated sign. The clock's illuminated hands were estimated to be visible on a clear night from as far away as twenty-four miles. Shortly after Jersey City's mayor set it going in May, Colgate placed advertisements in local papers announcing that the "Largest Clock and Largest Roof Sign in the World" was "now keeping correct time for the public."[44]

The Colgate clock surpassed Philadelphia's city hall clock as the largest in the world, and it retained that distinction even when, the next year, a twenty-five-foot, four-dial clock was mounted on New York City's Metropolitan Life Tower. These two companies exemplified the new corporate economy. The Colgate Company belonged to the new mass manufacturing sector that catered to the consumer market, branding its various lines of goods and promoting them through slick-paper magazine advertisements, newspaper promotions, billboards, and illuminated signs. Met Life, like other insurance companies, sought to cultivate trust and respectability. It needed to signify reliability, like banks, which had been installing clocks for at least a century. The Met Life building's clockworks also ran the company's offices, synchronizing clocks, bells, and gongs in the effort to coordinate the work and movements of its employees.[45] Through the twenties, corporate organizations sponsored gargantuan clocks, much as they sponsor sports arenas today. In 1926, a glass ball twenty feet in diameter nightly flashed the time each quarter hour from the top of New York City's Paramount Building, upon which hung twenty-five-foot clock dials.[46] The giant new commercial clocks were clearly associated with their sponsors—they were known as the Colgate clock, the Met Life clock, and the Paramount clock. Unadorned, they gained their distinction from their enormous size, which made them and their sponsors visible.

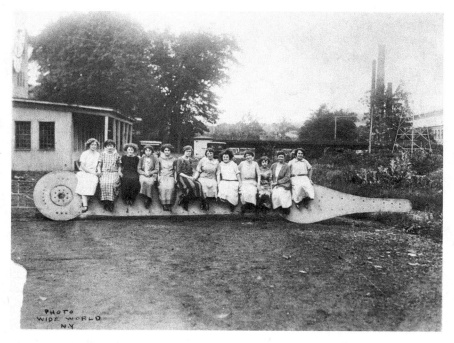

Hour hand of Colgate Company clock, 1909, Division of Technology, National Museum of American History, Smithsonian Institution. The Colgate clock's hour hand extended fifteen feet. The young women were probably employed in painting the clock hands. Note the clock hanging on the unidentified work shed to the left of the clock hand.

But there was an unease with these clocks, a distrust generated in part by their very size, by how they dominated the skyline, by how they towered over the streets, and in part by their inevitable malfunction. In the mid-1930s, for instance, fears accumulated that the corroded 1,600-pound hands of the Paramount clock "might plunge into" Forty-Fourth Street.[47] American studies scholar Nick Yablon traces apocalyptic visions of the Met Life clock's destructive power in period journalism and fiction. In one story titled "Darkness and Dawn," its "half-ton hands" fall to the street, leaving an enormous black hole. One newspaper report recounted the near-death experience of pedestrians who dodged masonry falling from the clock face. Less drastic were the accounts of New Yorkers' frustration when the Met Life's clock hands got stuck. In a 1919 piece of science fiction titled "The Runaway Skyscraper," the clock's hands run backward, a portent of the city's travel back in time. An illustration for this story shows two caricatured Native Americans standing in a tepee-covered field gazing up at the tower clock. Yablon names this illustration's temporal paradox of savagery and civilization "the future anterior," a verb tense in which the past occurs in the future. It is a concept that helps explain the attraction clocks laden with historical symbolism held for Americans. As Yablon explains, "Future anteriority offered an alternative perspective on the complex and obscure conditions of urban modernity. . . . Time unfolds in multiple directions, at multiple speeds, or in nonlinear patterns."[48] A clock's hands move forward always, except in fiction and occasionally in real life, but its decoration and setting can harken to the past, can connect the future with the past, can place the past in the future.

In 1900, little more than a year after Philadelphia's gargantuan pneumatic clock was set running, the chief of the Bureau of City Property unveiled a wooden "dummy clock" to the great consternation and disgust of the city. The gold-colored hands on its face "pointed permanently at 11:22 o'clock, the supposed time of the announcement of the Declaration of Independence."[49] The dummy clock helped to return the building to its original appearance, for it was shaped like an oversized grandfather's clock, resembling the two tall-case clocks that stood on the east and west walls of the building between the 1750s and 1820s. The question of whether the replica would run arose from political machinations surrounding the appointment in early 1900 of a new chief of the Bureau of City Property and new director of Independence Hall. Once the dummy clock was in place, the new appointees and all other public officials refused to take responsibility for it. The new chief of city property disingenuously explained: "I supposed the clock was to be a 'dummy,' as there was already a clock in the tower, never having heard that there was a place for an actual running clock." He further disclaimed: "There was nothing in the contract, so far as I knew or heard, to provide for a running clock."[50] Revelations, supported with documents, showed that the newly appointed chief instructed the contractor "that instead of having a real clock [at the west end], they would have a dummy." By choosing a dummy clock, he cut the budget for the clock from $18,000 to a little more than $1,000.[51]

The dummy clock that hung from Independence Hall was cast as a historical text, rather than as a cheap alternative to a running clock. But Philadelphia residents did not believe the spin. Newspapers reported that "the city [was] up in arms about the [state house] affair." Calling it "that terrible atrocity," some Philadelphians wrote to the papers suggesting that "the dial was a criminal waste of space."[52] According to one newspaper, the "incongruous and meaningless presence" of the dummy clock understandably excited a "storm of indignation and ridicule." The restoration committee members who assented to the budget-cutting plan to hang dummy clocks were accused of being "fakirs," "vandals," and in the business of advertising rather than of restoration.[53]

The outrage was not the result of being deprived access to the time. Throughout the controversy, the dials on the tower of Independence Hall continued to show the time, as did the countless other clocks and hour's bells in the city. Describing the dummy clock with its hands permanently set to 11:22 as "incongruous" and "meaningless" suggests that the diffusion of mechanical time took precedence over that of historic or national time. Philadelphians had come to depend on a network of public clocks as conveyances of clock time; within this context the Old State House held a venerable place. When and if city dwellers looked to it, they expected that it would show accurate time. That instead dummy clocks might show some fictitious moment when the Declaration of Independence was signed was insulting to intelligence and basic principles of timekeeping, not to mention patriotism.

Not that Americans wished to expunge national time from their clocks, as the popularity of monumental novelty clocks featuring American icons underscores. A legend about painted wooden clocks that served as jewelers' signs circulated around the nation for decades. The clock hands on these signs, which were also known as "dummy clocks," typically pointed to eighteen minutes after eight. Beginning in 1888, newspaper after newspaper printed the following answer to the question "Do you know why the hands of these dummy clocks invariably [approach] 8:20?" "Because it marks the exact moment when President Lincoln was assassinated in Ford's Theater." Reprinted through the second decade of the twentieth century, the story gained various twists, including one that has the custom originating in England where the hands marked the time of Queen Victoria's birth, another that holds that they commemorated the time of George Washington's birth, and still another that pinpoints the moment as being when the Declaration of Independence was completed.[54] Shortly after President William McKinley was assassinated in 1901, jewelers around the nation reinvigorated the legend that their clocks marked the time of Lincoln's assassination. Newspaper articles titled "New Hour of Fate," "Clock Hands to Be Changed," and "Clocks to Indicate Assassination Hour" described the plans of jewelers in Minneapolis, Denver, and Chicago to change the hands on their dummy clocks to five minutes before four, the moment when President McKinley was shot.[55] Arranging the hands in this manner left room on the sign for advertising.

During the public clock era, the dummy clock was but one way that timekeepers memorialized the dead. Some public clocks were designated as memorials for fallen soldiers or deceased city fathers. They could be modest in size—twelve inches across—or have much larger and more elaborate faces with letters spelling out the name of the memorialized.[56] Whether erected with subscription monies in honor of fallen soldiers or from the largesse of an individual seeking to memorialize a deceased spouse or child, memorial clocks are the ultimate memento mori. Across cultures, as the anthropologist Carol Greenhouse has shown, clocks have functioned symbolically as reminders of death, with their ceaseless ticking off of hours, days, and years.[57]

Memorial clocks and bells overlaid the reminder of death with a tribute to the well-lived life of a fallen hero, town father, or faithful wife. In 1887, Lawrence Graham of Whitewater, Wisconsin, presented a tower clock to the local congregational church "in the name of his departed wife" (who ironically went unnamed in the tribute).[58] The minister's sermon during the dedication services illustrates the multivocality of memorial clocks. His text, not surprisingly Ephesians "Redeem the Time," urged residents of Whitewater to make better use of their time on earth in preparation for the eternal time to be spent in salvation or damnation. The minister outlined a set of lessons that the clock could teach, which included, predictably enough, adjectives such as reliable, trustworthy, prompt, and steady. The central precept was that the clock should serve as a reminder that "time with us is fast passing, eternity is rapidly approaching." The minister then praised the clock for "minding its own business," noting "its sublime unconcern about the private affairs of others."[59] This is a curious rhetorical flourish, for elsewhere during this period the mechanical timepiece was portrayed as the ultimate busybody, driving people to nervous exhaustion, entering into their very psyches, and remaking them into watches or clocks.[60] But here the memorial clock placidly presides over a town of nosy people, improbably minding its own business. The minister draws a final lesson from the gift of the clock; like the clock, and the time it daily doles out, salvation too was a free gift: "Behold now

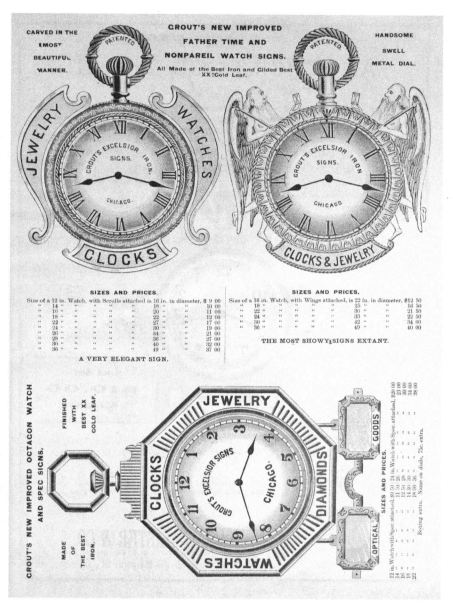

Listing for dummy clock in *Grout's New Improved Iron Signs*, B. F. Alister and Company catalog (Chicago, 1888), Hagley Museum and Library. Offered for sale are several different types of iron dummy clocks.

Robert E. Lee monument, Richmond, Virginia, ca. 1902, Detroit Publishing Company
Postcards, Photography Collection, Miriam and Ira D. Wallach Division of Art, Prints
and Photographs, New York Public Library. Richmond's Robert E. Lee monument has a
clock embedded in its central column. (Today it is still possible to buy a wall clock with
Robert E. Lee's likeness on the dial.) It is but one variation on the memorial clock, a
popular way to memorialize the dead during the public clock era.

is the accepted time, now is the day of salvation."[61] Twenty years later, a 1,580-pound bell put up on a library building in Mt. Airy, Pennsylvania, was inscribed with a similar admonition: "My ties are in Thy Hands. So teach us to number our days that we may apply our heart unto wisdom."[62] Clocks, then, as reminders of sacrificial death—of heroic soldiers, brave civil servants, and of Jesus the Savior—justified the symbolic death of the individual.

By the 1890s, the vogue for memorial clocks underscored the need of Americans for time to continue moving forward despite death. In 1894, J. T. Brown gave the Calvin Baptist Church in Hampton Falls, New Hampshire, a striking clock in memory of his wife, who he named in his gift. Gilded letters "Memorial Gift" took the place of the clock's numbers, and his wife's name, Ellen T. Brown, was engraved on the dial plate as well. Additionally, a brass plate made the intent of the gift quite clear: "This clock presented to the Town of Hampton Falls, May 30th, 1894 by John T. Brown of Newburyport, Mass. As a tribute to the memory of his Wife Ellen T. Brown." Memorial clocks were frequently found on the grounds of cemeteries, including Brooklyn's Greenwood Cemetery (1893) and Long Island's Pinewood Cemetery (1905). It was national news in 1897 when at its twenty-fifth reunion Harvard University's class of 1872 presented the university with a clock with four fifteen-foot wooden dials for the tower of Memorial Hall—a famous building devoted to the memory of Harvardians who fought in the Civil War.[63] A memorial clock and carillon erected in 1928 on the tower of the Louvain Library in Belgium in honor of 1,792 American engineers who died during World War I underscores the intensity of the need for clocks to announce an infinitude of time. The tribute proclaimed: "God's finger touched these men and they slept; but not the sleep of death, for in the peal of yon bells you may hear their immortal and triumphant song. . . . May victories for civilization and progress fill the years to come as the great hands of yonder clock slowly mark time's passage."[64]

Considering jewelers' dummy clocks and memorial clocks thus helps put Philadelphia's dummy clock episode in context and leads to several conclusions about public timekeeping at the turn of the century. First, daily timekeeping practices familiarized city residents with functioning timepieces; they could differentiate between the clock dials that showed the time, and those that did not. The only evidence that dummy clocks confused anyone is in the form of jokes that circulated about rubes from the countryside who mistook dummy clock time for *the time*.[65] Second, jewelers' efforts to capitalize in their signage on the exact fate-changing hour and minute for the nation points to how clocks were reshaping historical awareness. The idea that everything could turn in an instant was relatively new. Finally, the use of clocks as advertising by jewelers and others caused some discomfort, as we saw with the Milwaukee street clock incident discussed in chapter 5. Not only did it cross a line between public service (providing the time) and public exploitation (advertising), but more important, it drew attention to the fact that there were embedded interests behind all public clocks.[66] The Philadelphia replica's lack of authenticity further fueled disgust. It was not a clock; it was not sign in a jeweler's window; it was not advertising; it was not a memorial. It was instead a substitute for a real clock. So strong was public disapprobation that ten days after the dummy clock was hung, it was taken down.[67]

The paralyzed hands of Independence Hall's dummy clock presented national history as a fait accompli, something that happened at 11:22, rather than as an unfolding story in which the hands of time continued to move. The genius of claiming that jewelers' dummy clocks were set to the time of Lincoln's or McKinley's assassination was that they reinforced the sense that with the untimely death of a president, history—progress—was retarded. An epoch had ended. But the Declaration of Independence was an origin moment, and therefore it could not be represented on a dummy clock. The founding moment of the nation had to be remembered with clocks that ran, so as to convey the abiding sense that American history was continually unfolding, just as the hands on clocks were continually moving forward. One hundred dollar bills in circulation today show Independence Hall's clock tower with hands set to two and four; a Google search turns up conspiracy theories, in which the question is asked why "the clock is eternally stopped at 2:22?"[68]

Despite the mechanical failures attendant to Philadelphia's city hall clock and misguided restorationist sentiment about Independence Hall's clocks, their symbolic and utilitarian value could not be missed. Recall that when the city hall clock was set going for the first time, it was described as "an auspicious moment," indeed "one of the greatest events in the city's history."[69] The Independence Hall tower clock and hour's bell were said to represent the "auspicious moment" of the nation's birth, inheritors of the mantle carried by "the clock and bell that rung in the Rebellion and rang out the Rebellion."[70] They visually and aurally disseminated clock time well into the twentieth century. Indeed, the city council appropriated funds to assure that the state house clocks were accurate and even provided an extra appropriation in 1926, on the occasion of the sesquicentennial, to install an electric winding mechanism.[71] By radiating time across the city, clocks like these worked as public utilities, distributing a necessary resource, standardized time.

At the same time, veneration for the Liberty Bell grew to vast proportions, in part because it conveyed a sense of historic time. During the years

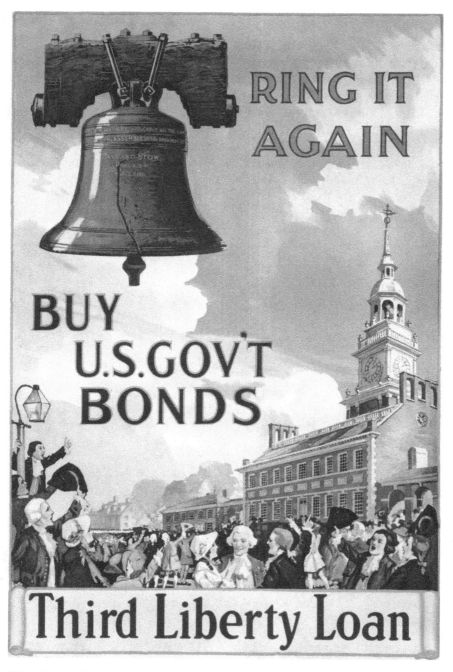

"Ring It Again!" Third Liberty Loan poster, 1917, in Willard and Dorothy Straight Collection, LC-USZC4-8886, Prints and Photographs Division, Library of Congress. Note the anachronistic appearance of the tower clock (installed in 1829) presiding over a scene meant to evoke 1776.

of the First World War, the Liberty Bell was used to sell "Liberty Bonds" and the war itself. It was tapped annually on the Fourth of July, New Year's Eve, and Memorial Day, emitting a faint sound. The bell on Independence Hall memorialized the dead; it tolled in tribute to two marines killed in Vera Cruz (1913), to Woodrow Wilson's first wife (1914), to Admiral Dewey (1917), and to Presidents Wilson (1924) and Coolidge (1933).[72] A cracked bell, dummy clocks, memorial clocks, monumental clocks, and monster timepieces—all were as important to how Americans marked modern times as public clocks, hour's bells, pocket watches, time services, and a single standard of time.

Content to Look at My Watch

The End of the Public Clock Era

In time perhaps I shall be content to look at my watch when I want to find out the time, but I am not now, it seems.

— *Brooklyn Eagle*, March 10, 1895

n 1895, a fire transformed the Brooklyn City Hall tower into "a charred mass." According to one report, the tower's clock faces and bell "had partly dropped through a hole in the roof." The morning after the fire, "people hurrying to business looked [in vain] from the elevated trains for a sight of the always friendly face of the city hall clock."[1] Several weeks later a journalist exclaimed about how profoundly "men of affairs" missed the clock: "There are hundreds of business men who note with regret its absence." While in the company of the reporter, one of these businessmen "looked across the street toward the city hall and smiled. 'There I've forgotten again,' he said. 'For years I have kept my appointments by the aid of the city hall clock and

I can't realize that it is gone.'" The informant went on: "Twenty times a day since that fire, I suppose, I look across." Trying to fashion an alternative to the public timepiece, the man of affairs mused: "In time perhaps I shall be content to look at my watch when I want to find out the time, but I am not now, it seems."[2]

Two months after the Brooklyn City Hall tower fire, the loss of the clock still smarted: a columnist likened it to "the amputation of a leg." Responding to a litany of complaints about the absence, in May 1895 the New York State legislature appropriated $100,000 to rebuild the tower complete with a new clock.[3] Throughout the following summer, Brooklyn was fraught with debate about architectural proposals to replace the tower. The "City Hall Tower Confab" as one headline read, could be characterized as "the high tower contingent" opposing the "low domers."[4] As "the war at city hall" raged, it was recalled that the old clock "was far more popular than any mayor whose official hours it timed. Folks were used to the tower, the clock and its announcement of 9 and noon. They just want[ed] to have it back again."[5] A year and a half later, it was reported that people were still "looking up at the sky to see what's o'clock." Purportedly, old timekeeping habits were hard to change: "[A] few . . . still unconsciously take their watches out of their pockets as they pass the city hall, and look up at the clock that isn't there."[6] Eventually, with the matter of the design resolved in favor of a cupola rather than a tower, Brooklyn's new city clock was installed in 1897, topped by a flag rather than the figure of Lady Justice that had been there for decades prior to the fire.

The interdependence of pocket watches and public clocks evident in Brooklyn in the mid-1890s began to deteriorate across the nation after the turn of the century. The expense of public timekeeping, always great, was called into question more frequently than in the nineteenth century. Recall that Philadelphia's chief of the Bureau of City Property tried to substitute a cheap dummy clock for a well-running one in 1900 largely as a cost-saving measure. Many other city officials around the country found themselves unwilling or unable to pay to regulate or replace civic timekeepers. In 1908, New York City's comptroller argued that the city was "too poor to spend $15,000 for a clock in the tower of the new Police Headquarters building." He called clocks "luxuries," and instructed the police commissioner to "keep time with his watch."[7] Watches were made to run with greater precision than they had fifty years earlier, so the necessity of constantly checking them against authoritative timekeepers had dimmed somewhat. But the real issue was that public clocks required vigilance. Their hands and faces were constantly subject to bad weather, corrosion, and the threat of fire. Their clockworks and striking mechanisms were susceptible to faulty design, malfunction, and poor

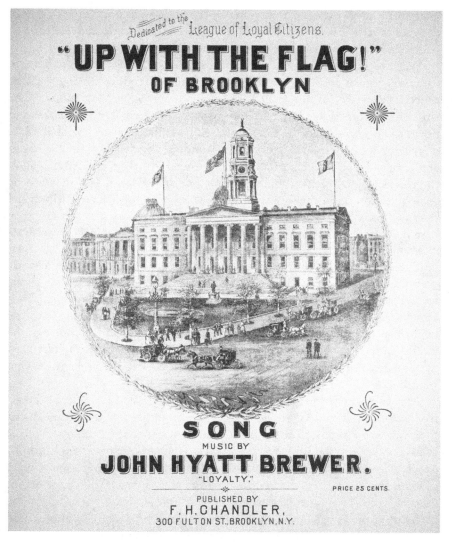

"Up with the Flag!," sheet music cover, 1895, Music Division, New York Public Library for the Performing Arts. Brooklyn, one of the largest cities in the United States before it was incorporated into Manhattan in 1898, showed a fealty to its town hall clock that indicates the degree to which modern time discipline was rooted in an admixture of public clocks, pocket watches, and schedules adhering to standard time.

maintenance.[8] Worst of all, regulating clocks to show accurate time was more than just a hassle, despite the US Naval Observatory's cooperative efforts to telegraph its time signals, it was often impossible.

Even in the 1920s, the absence of a single source for accurate time in America's "great cities" was a source of journalistic fodder. Taking what he called, "a 'sample' of time as it is known to the average New Yorker," one curious reporter "walked up one side [of a 'typical street'] several blocks and down the other, noting the time displayed on outside clocks maintained by merchants for pedestrians' convenience, inside clocks visible from the street, and jewelers' clocks and chronometers in windows." He marveled that all of the clocks he saw "were running merrily and being taken as a standard by somebody, although they were as much as forty minutes apart!" Upon the basis of this lack of consistency among New York's clocks, the reporter suggested distributing "the correct time over electric light systems" as was done in Montevideo, Uruguay. Every night at eight in that South American capital, the reporter recalled from a recent visit, the government-owned electric company dimmed the lights, providing an accurate and reliable time signal for the adjustment of watches and household clocks. He speculated that if adopted in the United States, this simple service would reach millions.[9] Electric companies never signed on to a plan like this one; but over the course of the 1920s, other ways of synchronizing clocks, some dependent on electricity, did emerge, rendering public clocks with all their imperfections somewhat superfluous.

The wireless transmission of time signals, inaugurated in 1903, also contributed to the demise of public clocks. The adoption of wireless time distribution was, inevitably, slow, partially because public uses for radio were hard at first to fathom, though navigators quickly incorporated wireless transmission into their tool kits. Wireless hastened "the globalization of clock time," as social theorist Barbara Adam explains. She argues that 1913 "is the beginning of world time," because when the first time signal was sent across the globe from the Eiffel Tower the morning of July 1 "it established one time for all."[10] The US Naval Observatory sponsored radio time service from Fort Meyers (near Arlington, Virginia) that same year, and by 1915 the service was available from eight radio stations.[11] Leaving aside the sticky details of whether there is even today "one time for all," it is nevertheless the case that the effect of wireless on timekeeping was profound. It brought to an end the public clock era, and it, as Adam explains, "facilitated instantaneity and simultaneity."[12]

Although the public clock era is over, the importance of the clock time it instantiated persists. The time stamp is a fundamental feature of modern life.

Modern communications media, particularly radio, television, and the Internet, depend on it: clock time is embedded in how each marks its products. While electronic media do not need public clocks, their various iterations depend on the conceit of clock time. Tempting as it is to argue, as some do, that computers represent a new way to reckon and order time, the fact is that they are built upon a grid of clock time that was experimented with and developed over the course of several centuries.[13]

When commercial radio broadcasts were first being aired, Christine Frederick, a leading home economist during the Progressive Era, commissioned a study about possible domestic uses for radios. In addition to "entertainment at mealtime and teatime," her 1922 study investigated "uses of radio for time and information."[14] Under a pseudonym, Frederick photographed possible scenarios in which the radio figured; in two "time and information" photographs, elderly men with pocket watches in hand sit beneath regulators, a type of clock usually found in offices, not homes. But no matter, the point was that the radio could bring clocks into agreement. It is not unusual at all that Frederick, who had been applying Frederick Winslow Taylor's scientific management principles to household tasks since her 1912 publication *The New Housekeeping: Efficiency Studies in Home Management,* had time on the mind when she imagined domestic uses for the radio. Little did she realize the degree to which radio and telephone would synchronize people with their timekeepers, allowing them to achieve degrees of time efficiency unimaginable in the 1920s.[15]

Several decades earlier, the chemist and student of astronomy Francis G. DuPont sought out a similar degree of synchronicity, though without the benefit of wireless transmission or the radio. In late 1880s, DuPont oversaw the design and construction of the world's largest mills for the manufacture of black blasting powder, and in the early 1890s, he compounded smokeless gunpowder. Ever since the introduction of national standard time in 1883, DuPont had been obsessed with acquiring accurate time and synchronizing the clocks in his Delaware factory and adjacent residence, where he had, by his own description, "a neat little observatory, and a transit instrument, and also a good clock."[16] In 1890, he built a circuit that connected a regulator to six dials hanging in his house. A number of other clock dials were on the circuit as well: in a stable two hundred feet from his house; in his factory, which was about two thousand feet from his house; in a building sheltering an electric dynamo; in several rooms of another residence; in the watchman's house; and in a machine shop. In all, the regulator was connected to sixteen

Man setting pocket watch and household clock to radio time, 1922, Phyllis Frederick Photo Service, Schlesinger Library, Radcliffe Institute for Advanced Studies, Harvard University. In this awkward photograph, the contours of modern time discipline are evident—a radio, a pocket watch, and a reliable clock. Headphones were optional.

clock dials spread over an expanse of about two miles. The DuPont inventor exclaimed about the time circuit: "Really the whole thing from the little dynamo to the Regulator gives the greatest satisfaction."[17]

Over the next decade, DuPont enthusiastically corresponded with the Standard Electric Time Company, which sold him parts for what he described as "an excellent time system."[18] After electric power became widely available in the 1880s, inventors and entrepreneurs, like those behind the Standard Electric Time Company, initiated experiments on how to electrify timekeeping. Two innovations transformed timekeeping. One was the design for a self-winding clock, harnessing electricity to wind weight-driven clocks. The other innovation connected secondary clocks to a master clock using electric circuits, creating the conditions under which clocks could be effectively synchronized.[19] Electrical timekeeping was the future, though in the 1880s and 1890s, only a handful of companies in addition to the Standard Electric Time Company were in the business.

Francis G. DuPont seated in his office, ca. 1890, Hagley Museum and Library. An inventor and engineer, DuPont was obsessed with clocks, spending much of his spare time tinkering with them.

After 1900, interest in electrical timekeeping grew, as did the number of companies offering time systems to managers of large offices and institutions like schools, as well as marketing electrically wound clocks (often called "self-winding") to householders.[20] In the 1920s, innovations in clock design and in electric power distribution together resulted in precision clocks that did not need weights or springs to control their rate, but instead depended on the motion of alternating current to mete out the time.[21] But until the electric power supply was reliable, and the design of clocks moved away from three centuries' worth of investment in weight-driven and spring-driven movements, electricity was mostly called upon to wind or synchronize clocks. Despite electricity's transformation of timekeeping, clock winders still visited thousands of households and hundreds of public clocks every week in all the major American cities well through the 1940s.[22] DuPont discovered the Standard Electric Time Company in 1890, several years after it entered into business; in the vanguard, he used its system and parts to perfect the time circuit he had built to synchronize the clocks in his home, factory, and office.

Despite being driven one fall night in 1897 to call for an axe because he "felt like smashing the clock [his regulator] to pieces," in 1899 DuPont wrote, "I am happy to say that after about nine years of use I find it as near perfect as a time system can be."[23] He delighted in the fact that the dials ran "in inaccessible places without any attention." Boasting, DuPont wrote, "I have dials in many of the rooms in my house." But it was about the one in his bedroom that he exulted at great length: "I have a dial on the wall of my bed room, in such a position as to be easily seen from the bed. The dial is transparent, and has four electric lights behind it." DuPont had rigged up a "push button at the end of a flexible cord" hanging near his pillow, so that "at any moment" he could "readily see the time of night without moving, or being disturbed." He was not yet finished with his reverie: "You can not imagine the comfort, if one is awakened at night, of pressing the button and seeing that, say, it is three o'clock."[24]

Henry Ford, like fellow industrialist DuPont, was obsessed with mechanical timekeepers. Like home economist Frederick, Ford was driven to implement time efficiency wherever he could. He was given his first watch in 1875 at the age of twelve. After Ford's mother died two years later he wrote, "Now home is like a watch without a mainspring." Shortly after the introduction of the national standard time system, Ford put together a watch with two dials; one dial showed local time, and the other standard time. By the end of his life, he had a collection of more than twelve hundred complete watches, and another thousand incomplete ones, many of which were given to him by the managers of America's great first great watch manufacturers, Waltham and Elgin. Watch repair was the automobile maker's first engineering feat: Ford recounted that after Sunday school one day he repaired a friend's watch with a tool he made by grinding "a shingle nail into a screwdriver." With this crude implement, he got the cap off the watch, saw what was wrong with it, fixed it, and consequently "got quite a reputation." Thereafter, Ford frequented jewelry stores in Detroit looking for watch parts and tools, and he even took a part-time job as a watchman at the McGill Jewelry Shop. Although he sketched out plans for watchmaking machinery, he decided that watch manufacture was the wrong path to success, since, as he put it, watches were "not universal necessities."[25] Ford was quite right, pocket watches would go the way of the buggy and the whip, and maybe today so are wristwatches, but clock time had lasting power.

The economic crisis of the 1930s generated a discourse of time arrested, of regression, of decay. But clock time maintained its salience as an indicator of national strength and modern values. The penetration of radio time and electricity into most crevices of American life by the end of the decade rendered obsolete or just quaint the drop of time balls, the nine o'clock flash of lights, and the ringing of the hours. Other techniques and apparatuses for disseminating the time, particularly centrally located, prominent clocks, persisted. In 1930, General Time Instruments Corporation, a general holding company, acquired the Seth Thomas Clock Company; and in 1933, E. Howard's clock business, which had had several near collapses, reorganized as Howard Clock Products Company. During the 1930s, some clock towers on local courthouses, city halls, and other downtown buildings were taken down, some with public works funds footing the bill. Others were refurbished. In the early 1940s, still more public clocks, particularly street clocks, were removed and disassembled in order to contribute to the wartime drive for scrap metal. The public clock era was over.

Although wristwatches were widely available by the 1930s, many Americans continued to use pocket watches, perhaps in part because tightened purse strings precluded upgrading.[26] Since the teens and twenties, military men and the fashion conscious gravitated toward wristwatches, whose lines included Depollier's "DD Khaki Watch," Waltham's "The Cadet," Hamilton's "Gents Wrist Watch," Dueber Hampden's "Man O'Fashion" and "Play Boy," and Elgin's Parisienne series for women (which featured the "Madame Agnes" and the "Madame Jenny").[27] A 1921 Waltham catalog protests too much: "The Strap Watch is no longer a fad or an eccentricity." Slow sales seemed to contradict the pitch's claim that "it is an essential of everyday life—a convenience that men are appreciating more and more."[28] Before the end of the twenties it was clear that pocket watches were indeed on their

Russell Lee, *Clock in the Home of John Landers,* January 1937, US Resettlement Administration, LC-USF33-011144-M2, Prints and Photographs Division, Library of Congress. Russell Lee was attentive to the iconic power of personal timepieces. In his photography for New Deal agencies, he provided shots of Americans attending closely to clock time. The unsynchronized timepieces in this farmer's house near Marseilles, Illinois, convey an air of time standing still. A photograph cannot indicate whether a timekeeper's hands have stopped, but in many from the Great Depression it appears as though they have.

way out, not just for fashion's sake, but for timekeeping's as well. Wristwatches were easier to consult. The lifetime of pocket watches may have been extended by the economic depression of the 1930s, but their disposal was hastened by the crisis as well. As had been the custom since the 1700s whenever hard times hit, pocket watches were sold or pawned, their utility as investment outweighing whatever timekeeping benefits they might have offered. And millions of others were stowed away when they needed repairs, becoming over time sentimental reminders of what seemed to be simpler times. By 1935, 85 percent of the watches made in American factories were wristwatches, a dramatic increase from 15 percent in 1920.[29] Demand for wristwatches would swell in the 1940s and 1950s. They, along with synchronous electric clocks, telephone time services, and wireless communications facilitated modern time discipline.

In the 1950s, a new instrument of time discipline and type of government clock debuted, the countdown clock. The practice of counting down the time can be partially traced to rituals associated with ushering in the New Year. It was also part of mining, the building trades, and waging war,

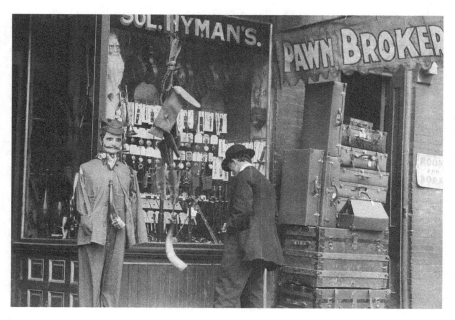

Lewis Hine, *Looking at the Guns in the Local Pawn Shop*, 1930, Photography Collection, Miriam and Ira D. Wallach Division of Art, Prints and Photographs, New York Public Library. Dozens of pawned pocket watches are visible in the window of a Nashville pawnshop just a year after the stock market crash of 1929.

wherein dynamite or cannon effected great destruction. Special clocks were designed as a time-management procedure to control the launch of rockets and missiles. They were widely adopted, found on game shows, embedded in the televised broadcasts of football and basketball games, and trotted out to measure and control anticipation. Near the end of the twentieth century, the Chinese installed a prominent countdown clock in Tiananmen Square; it marked the time until Hong Kong's return to China. This clock, on a panel a thousand feet high and six hundred feet wide, illustrates yet another way clock time has been deployed in the service of nation building.[30] Or consider the countdown clocks that appeared as the US Congress tried to avoid a government shutdown in April 2011. Various television and web news outlets presented countdown clocks, some of which were calibrated to a tenth of a second, which visually "conveyed the pressure the legislators were under" to come to an agreement about servicing the national debt. As one journalist observed, "The clocks added a dose of visual drama to what was otherwise

Digital countdown clock and analog military-time clock, Vandenberg Air Force Base, Space Launch Complex 3-West, ca. 1963, HAER CAL, 42-LOMP, 1A–33, Prints and Photographs Division, Library of Congress. Note the many clocks, including a countdown clock, in this missile launch operation room.

a visually uninteresting day of press statements and floor speeches." At one point CNN enlarged its clock's numbers so they were as big as the anchors' heads. The countdown clocks were the stuff of jokes, satirized for a national audience by comedians like Stephen Colbert and others.[31]

While the US government still maintains a vast array of clocks in the buildings of its many agencies in Washington, DC, and throughout the nation, it also provides accurate time through a number of powerful coordinating technologies, mostly overseen by the National Institute of Standards and Technology (NIST), an agency of the Department of Commerce. Before the agency that would become NIST moved to the National Bureau of Standards in 1901, it was part of the Office of Weights and Measures. Between 1901 and 1941, it tested timepieces using two specialized pendulum clocks, each made in Europe. After experiments at Bell Laboratories and in England during the 1920s established the precision timekeeping qualities of quartz, NIST began to experiment with quartz oscillating mechanisms. In the 1960s, quartz transformed personal timekeeping; it increased the precision timekeeping of wristwatches and alarm clocks while decreasing the price of each more significantly than any other horological innovation.[32]

In 1949, an atomic clock was developed in NIST's laboratories; on the basis of the atomic clock, an international agreement in 1967 redefined the second as "the duration of 9 192 631 770 periods of the radiation corresponding to the transition between the two hyperfine levels of the ground state of the cesium-133 atom."[33] Time specialists were no longer astronomers using instruments to keep track of the earth's movement, but physicists. NIST and the US Naval Observatory together set coordinated universal time (UTC), often referred to as "world time." UTC is accessible in a variety of ways; most of us probably receive it through our networked cellular phones and computers.[34] NIST also distributes this time through various media: two shortwave radio stations (WWV near Fort Collins, Colorado, and WWVH near Kauai, Hawaii), an Internet time service, an automated computer time service, a telephone time service, and a US government web clock (http://time.gov/).[35]

Until 2004, a shortwave radio signal from the WVV station sent the correct time to the master clock of New York City's Grand Central Terminal. It then corrected the station's twenty clocks over an electric circuit. The electromechanical system did not work especially well, and the clock time was not synchronized with the time shown on the video monitors throughout the vast transportation hub. Few of the 700,000 people who pass through Grand Central every day consider the public time as accurate, preferring, as one commuter said, to "just use the clock on my phone." Even the train conductors consulted their own wristwatches, rather than platform clocks, when deciding to leave. Aware of the lack of synchrony, the managers in charge of Grand Central's time installed a new master clock connected by satellite to NIST's atomic clock in Boulder, which would keep all the station's clocks accurate twenty-four hours a day.[36] Contests for authority over the time had long ago ceased, but the quest for accuracy and synchrony as evident in the effort to regulate Grand Central's clocks marks our times as much as it did the public clock era.

I began this book with jewelers and end with railroad station managers. I do so not to undercut the importance of civic officials and institutions, but to highlight the partnerships and cooperation crucial to modern timekeeping and modern time discipline. No single entity could have made clock time one of the basic building blocks of modern life and societies, but together a variety of private and public partnerships did. Of fundamental importance was the willingness, indeed the eagerness, of Americans to acquire and use personal timekeepers that connected them to the network of clock time as shown on public clocks and rung on bells. They persisted in their attempts at synchronicity even when clocks and watches proved unreliable. Eventually the technology caught up with, and perhaps surpassed, their aspirations.

NOTES

ABBREVIATIONS

The following abbreviations refer to archives, special collections, scholarly journals, and newspapers cited in the notes.

AHR: *American Historical Review*

Allen Papers: William F. Allen Papers, New York Public Library

Archives Center, NMAH: Archives Center, National Museum of American History, Smithsonian Institution, Washington, DC

Arthur Collection: James Arthur Collection, Clippings file, Archives Center, National Museum of American History, Smithsonian Institution, Washington, DC

Baker: Baker Library Historical Collections, Harvard Business School

BCP-IH: City of Philadelphia, Bureau of City Property, Independence Hall Division, Records, 1873–1951

BE: *Brooklyn Eagle* and *Brooklyn Daily Eagle*

Bond Records: Records of William Bond and Son, 1763–1921, Harvard Historical Scientific Instruments Collection, Harvard University

BPL: Boston Public Library

CDN: *Chicago Daily News*

DL-SMU: DeGoyler Library, Southern Methodist University

DuPont Clock Correspondence: DuPont Family Miscellany, accession 1421, Hagley Library, folder 3, "Correspondence between F. G. DuPont and the Standard Electric Company"

DuPont Papers: Papers of Francis Gurney DuPont, accession 504, Hagley Library

FLP: Free Library of Philadelphia

HC: E. Howard Collection, Archives Center, National Museum of American History, Smithsonian Institution, Washington, DC

HSP: Historical Society of Philadelphia

IHA: Archives, Independence Hall National Historical Park

IH-CCF: Independence Hall Clock Card file, Archives, Independence Hall National Historical Park

JEH: *Journal of Economic History*

Landauer Collection: Bella C. Landauer Collection of Business and Advertising Ephemera, PR03, New-York Historical Society

LC-PPD: Library of Congress Prints and Photographs Division

LCP: Library Company of Philadelphia

Luce Center: Henry Luce III Center, New-York Historical Society

NA: Philadelphia North American

NAWCC: National Association of Watch and Clock Collectors, Columbia, Pennsylvania

NB: Bulletin of the National Association of Watch and Clock Collectors

NDSC: University of Notre Dame, Rare Books and Special Collections, Manuscripts of the American Civil War

NMAH-SI: National Museum of American History, Smithsonian Institution, Washington, DC

NMIH: National Museum of Independence Hall

NYHS: New-York Historical Society

NYPL: New York Public Library, Astor, Lenox and Tilden Foundations

NYPL-PA: New York Public Library for the Performing Arts, Astor, Lenox and Tilden Foundations

NYT: New York Times

PCA: Philadelphia City Archives

Perkins Scrapbook: Samuel C. Perkins, Scrapbooks, Philadelphia Public Buildings, vol. 29, Historical Society of Philadelphia

PL: Philadelphia Public Ledger

PP: Philadelphia Press

PR: Philadelphia Record

PT: Philadelphia Times

PTel: Philadelphia Telegraph

Public Property Scrapbooks: Public Property Scrapbooks, 1894–1901, Philadelphia City Archives

ST: Seth Thomas Company account books, held by the Library of the National Association of Watch and Clock Collectors, Columbia, Pennsylvania

USHO: United States Hydrographic Office

USHO-Records: Records of the United States Hydrographic Office, United States Navy, National Archives Records Administration, Record Group 37

USHO-Correspondence: Correspondence of the United States Hydrographic Office, 1908–27, National Archives Records Administration, Record Group 37, entry 41, boxes 49–51

USNO: United States Naval Observatory

USNO-Records: Records of the United States Naval Observatory, National Archives Records Administration, Record Group 78

Valley of the Shadow: Valley of the Shadow, Two Communities in the American Civil War, Virginia Center for Digital History, University of Virginia, http://valley.lib.virginia.edu/

Waltham Records: Waltham Watch Company Records, Treasurers Reports, 1859–79, boxes AD-1 and AD-2, Baker Library Historical Collections, Harvard Business School

Warshaw-WC: Warshaw Collection of Business Americana, Watch and Clockworks, Archives Center, National Museum of American History, Smithsonian Institution, Washington, DC

INTRODUCTION

1. Joachim Giaver and Frederick Dinkleberg's Jewelers Building is located at 35 East Wacker Drive. Shortly after it was completed in 1927, it was unofficially renamed the

Pure Oil Building because the Pure Oil Company was its largest tenant. It should not be confused with another Chicago architectural landmark, Adler and Sullivan's Jewelers Building, 15–19 South Wabash Avenue. See Timothy N. Wittman, "35 East Wacker Drive Building," Preliminary Staff Summary of Information Submitted to Commission of Chicago Landmarks (December 1992), typescript report in author's files; Stephen Sennott, "Chicago Architects and the Automobile, 1906–1926," in *Roadside America: The Automobile in Design and Culture*, ed. Jan Jennings (Ames: Iowa State University for the Society of Commercial Archaeology, 1990), 157–78. See also Al Chase, "Jewelers' New $10,000,000 Bldg. Partly Garage," *Chicago Daily Tribune*, April 16, 1924; Ernest Fuller, "Famous Chicago Buildings," *Chicago Daily Tribune*, April 11, 1959.

2. About the state's claim to own time, see Carol J. Greenhouse, *A Moment's Notice: Time Politics across Cultures* (Ithaca: Cornell University Press, 1996); Gerhard Dohrn-van Rossum, *History of the Hour: Clocks and Modern Temporal Orders*, trans. Thomas Dunlap (1992; reprint, Chicago: University of Chicago Press, 1996); Michael Adas, *Machines as the Measure of Men: Science, Technology, and Ideologies of Western Dominance* (Ithaca: Cornell University Press, 1989). Historian Robert Rotenberg calls the civic organization of schedules "an extraordinary instrument of power and domination" (*Time and Order in Metropolitan Vienna: A Seizure of Schedules* [Washington, DC: Smithsonian Institution Press, 1992], 2). In a close study of the formation of crowds in Britain between 1790 and 1835, the historian Mark Harrison also underscores the role of civic power over time and schedules ("The Ordering of the Urban Environment: Time, Work and the Occurrence of Crowds, 1790–1835," *Past & Present* 110 [February 1986]: 134–68).

3. Masha Green, "Medvedev's Time," *Latitude* (blog), *NYT*, November 3, 2011, http://latitude.blogs.nytimes.com/2011/11/03/medvedevs-time/ (accessed March 11, 2012); Corey Flintoff, "In Russia, a Debate over How to Set the Clock," *National Public Radio*, February 14, 2012 http://www.npr.org/2012/02/14/146875857/in-russia-a-debate-over-how-to-set-the-clock (accessed February 15, 2012).

4. Paul Glennie and Nigel Thrift, *Shaping the Day: A History of Timekeeping in England and Wales, 1300–1800* (New York: Oxford University Press, 2009). Glennie and Thrift explain that clock time results from a confluence of "*concepts, devices,* and *practices*" (9, emphasis in original). They argue that clock time enhanced efforts to regulate and coordinate social action, but they dispute its centrality in disciplining social actors (43, 45–47). David D. Hall's rich reading of the diary of seventeenth-century New Englander Samuel Sewall provides a good example of how clock time took shape as a set of concepts, devices, and practices. Sewall recorded the time of nearly every event in his diary, while also attending to the variegated measures of time's passages and meaning ("The Mental World of Samuel Sewall," chap. 5 in *Worlds of Wonder, Days of Judgment: Popular Religious Belief in Early New England* [New York: Knopf, 1989], 213–38).

5. Michael J. Sauter, "Clockwatchers and Stargazers: Time Discipline in Early Modern Berlin," *AHR* 112 (June 2007): 685–709, esp. 706–7.

6. Robert Hassan and Ronald E. Purser, introduction to *24/7: Time and Temporality in the Network Society*, ed. Robert Hassan and Ronald E. Purser (Stanford: Stanford Business Books, 2007), 1–21. Self-identified "theorists of time," Hassan and Purser explain: "The meter of clock time that drove the industrial revolution is now being compressed and accelerated by the infinitely more rapid time-loaded functions of high-speed computerization." They further observe: "Humans, through their ever-increasing use of networked PCs, PDAs, cell phones, voicemail, faxes, pagers, computer games, and so on, are creating an accelerated temporal ecology (an experience of time) that is entirely unprecedented" (11). Barbara Adam in her foreword to Hassan's and Purser's *24/7* describes "Informa-

tion Communication Technology (ICT) temporality" as being "globally networked" (in contrast with clock time which is "globally zoned"), instantaneous, simultaneous, and "chronoscopic" (in contrast to clock time, which is "spatially constituted") (x–xi).

7. There are many historical accounts about timekeeping before mechanical clocks. See especially Anthony Aveni, *Empires of Time: Calendars, Clocks, and Cultures*, rev. ed. (Boulder: University Press of Colorado Press, 2002); Melissa Barden Dowling, "A Time for Caesar: The Julian Calendar and Roman Politics," in *Making History: Essays in Honor of Ruth Sharp Altshuler* (Dallas: Southern Methodist University, 2007): 87–115; Joseph Needham, Wang Ling, Derek J. deSolla Price, *Heavenly Clockwork: The Great Astronomical Clocks of Medieval China*, 2nd ed. (Cambridge: Cambridge University Press, 1986); Wu Hung, "Monumentality of Time: Giant Clocks, the Drum Tower, the Clock Tower," in *Monuments and Memory, Made and Unmade*, ed. Robert S. Nelson and Margaret Olin (Chicago: University of Chicago Press, 2003), 107–32.

8. Glennie and Thrift, *Shaping the Day*, 91.

9. Dohrn-van Rossum, *History of the Hour*, 171. Dohrn-van Rossum suggests that "the possession and use of clocks are indicators of modernity" (3). About clocks and bells in Spanish colonies, see Charles Gibson, *Tlaxcala in the Sixteenth Century* (Palo Alto: Stanford University Press, 1967), 147; Jay Kinsbruner, *The Colonial Spanish-American City: Urban Life in the Age of Atlantic Capitalism* (Austin: University of Texas Press, 2005), 142n3. Kinsbruner notes that "the Spanish Crown from an early date wanted at least one public clock in the colonial urban centers."

10. Dohrn-van Rossum, *History of the Hour*, 128. The Italian poet Petrarch first used the term "public clock." Dohrn-van Rossum defines a clock as a mechanism that "indicate[s] visually or aurally the sequence of hours of the full day. The kind of technology used—tower warden clock or tower clock—is not important." He defines public as "any large number of people, especially city-dwellers, but also members of a princely residence, the neighbors of a monastery, or the members of a university." Finally, Dohrn-van Rossum stipulates that to consider a clock "public," contemporaries ought to have referred to it as "communal" or "public" (129). About the history of clocks more generally, see especially Davis Landes, *A Revolution in Time: Clocks and the Making of the Modern World*, rev. ed. (Cambridge: Harvard University Press, 2000).

11. Frederick M. Shelley, *Early American Tower Clocks: Surviving American Tower Clocks from 1726 to 1870* (Columbia, PA: National Association of Watch and Clock Collectors [NAWCC], 1999). Donn Haven Lathrop's website, American Tower Clocks, 1717–2011, http://homepages.sover.net/~donnl/amtc.html (accessed February 7, 2012), is a useful compilation of secondary and primary sources related to the installation of tower clocks in the United States. About public clocks in one American city, see Chris DeSantis, with photographer Vinit Parmer, *Clocks of New York: An Illustrated History* (New York: MacFarland, 2006). Back issues of the *Bulletin of the National Association of Watch and Clock Collectors* (*NB*) contain articles concerning public clocks, especially articles devoted to the restoration of tower clocks. See, for example, Henry F. R. Watts, "Tower Clocks: A Survey of Many Interesting New England Specimens Made before 1850," *NB* 4 (1950): 94–102.

The latest example of the quixotic goal of counting clocks is the NAWCC's "Public Time Initiative." This project started in 2004 to provide an updated inventory of public clocks in the United States; at present it is in its "Spot-a-Clock" phase in which information about extant public clocks is solicited from NAWCC members. See Frank del Greco, "The Public Time Initiative: The Idea of a Decade," *NB* 46 (2004): 291–92; "Spot-A-

Clock," on the NAWCC website, http://www.nawcc.org/index.php/resource-center/ spot-a-clock (accessed May 12 and 17, 2011). As of the summer of 2011, only 502 entries had been made in the Spot-A-Clock database. Another initiative to inventory the extant public clocks in the United States is led by the organization Save America's Clocks, http:// www.clocks.org/ (accessed November 14, 2011). Founded in 1996, it seems to have foundered. At the end of 2011, its website listed fewer than one hundred clocks.

12. The sale of 15,000 public clocks between 1871 and 1911 is an estimate based on the extant account books of the E. Howard Clock Company and the Seth Thomas Clock Company. As I explain below, I estimate that E. Howard sold 12,971 public clocks between 1871 and 1911 and that Seth Thomas sold 2,000 tower and street clocks, coming to a total of 14,971, which I have taken the liberty of rounding up to 15,000. Seth Thomas sold considerably fewer public clocks than did E. Howard, largely because the company did not sell small interior clocks meant for public spaces or master-secondary clock systems, as did E. Howard. Seth Thomas did market tall-case (grandfather) clocks, but its records for such sales are not extant and thus are not included in this estimate.

Between 1871 and 1905, 8,149 separate entries appear in the E. Howard books. Analysis of 31 percent of the entries (2,550) shows that it is probable that a third of all the orders were for repairs or for watch clocks (an early form of a time clock). The remaining two-thirds, or 5,433 orders, likely were for public clocks. Many of the orders for public clocks were for more than one clock, in some cases dozens of clocks. A conservative estimate is that 100 orders out of the 5,433 orders were for 2 clocks, totaling 200. Of these orders, 500 likely were for a dozen clocks, coming to 6,000 clocks. If the remaining 4,833 orders were for a single clock, then E. Howard sold a total of 11,033 public clocks between 1871 and 1905, or about 323 clocks a year. In order to estimate E. Howard's sales through 1911 (so as to bring the data into alignment with the Seth Thomas data), I conservatively estimated that the company sold the same number of clocks per year as it had on average between 1871 and 1905: 323 clocks per year for six years come to an additional 1,938 clocks. Thus, I estimate that E. Howard sold 12,971 public clocks 1871–1911.

In the account books of Seth Thomas, 1,124 orders for clocks were recorded for 1877–93 and 1903–11, or about 50 clocks per recorded year. Nearly all of these orders were for a single clock movement, albeit movements large enough to move the hands of four and sometimes even eight dials. Since there are not extant records for each of the years, I estimate that Seth Thomas sold 50 clocks a year, for a total of 2,000 clocks between 1871 and 1911.

The Archives Center at the Smithsonian's National Museum of American History (NMAH), holds the E. Howard Collection (HC), which contains all of the extant E. Howard account books. The NAWCC has digitized the E. Howard account books for the use of the association's members; they are available on the NAWCC's website. Books containing orders for the entire array of clocks produced by the company cover the periods April 22, 1871–January 27, 1886; February 1, 1888–April 1, 1891; November 14, 1892–January 30, 1899 (box 8, vols. 1–6; box 9, vols. 1–3; box 10, vols. 1–3). The boxes are not chronologically organized, though the individual volumes are. An order book for tower clocks covers the period between January 1902 and December 1905 (box 11, vol. 1) and another book of watch clock orders extends from January 1904 to October 1918 (box 11, vol. 2). Additionally, a lengthy undated customers' list in the form of an index is available (box 12, vols. 3 and 4).

The Seth Thomas Company account books (ST) held in the NAWCC Library detail "Tower Clock Installations" for April 1877–July 1877, September 1877–April 1883, June

1884–October 1893, May 1903–July 1911. There is an incomplete index by town of all tower clock installations between 1872 and 1942. Because they are piecemeal and incomplete, I did not consult six account books that cover the years between 1922 and 1942 (books Q–U and W), which are held by the NAWCC Library. My thanks to Christina Zienkowsky who assisted me in building the Seth Thomas database covering the years between 1906 and 1911. The NAWCC has digitized the Seth Thomas account books for the use of the association's members; they are available on the NAWCC website.

13. Joseph G. Baier, "Mathias Schwalbach," *NB* 15 (August 1973): 1218; Jack Linahan, "Nels Johnson, Michigan Clockmaker," *NB* 48 (August 2006): 391–402. The appendix of Shelley's *Early American Tower Clocks* lists the names and locations of tower clock makers working in the United States during the eighteenth and nineteenth centuries. In the eighteenth century, Shelley identifies twenty-four tower clock makers working in the United States: nine in the Middle Atlantic, thirteen in New England, and two in the South. Over the course of the nineteenth century, he identifies eighty-eight tower clock makers: thirty-three working in the Middle Atlantic, thirty-four working in New England, eight in the South, and thirteen in the Midwest and Far West. Most of these clockmakers made just a few tower clocks.

14. About electric clocks in the United States, see Michael O'Malley, *Keeping Watch: A History of American Time* (Washington, DC: Smithsonian Institution Press, 1990), 156–57; Ian R. Bartky, *Selling the True Time: Nineteenth-Century Timekeeping in America* (Stanford: Stanford University Press, 2000), 30, 52–53, 56, 175; Carlene Stephens, *On Time: How America Has Learned to Live by the Clock* (Boston: Bulfinch Press, 2002), 35, 128, 141, 148–49, 191, 194–95. More research is needed into electrical timekeeping in the United States.

About electrical timekeeping in Britain, see Hannah Gay, "Clock Synchrony, Time Distribution and Electrical Timekeeping in Britain, 1880–1925," *Past & Present* 181 (November 2003): 107–40; David Rooney and James Nye, "'Greenwich Observatory Time for the Public Benefit': Standard Time and Victorian Networks of Regulation," *British Journal for the History of Science* 42 (March 2009): 5–30; Iwan Rhys Morus, "'The Nervous System of Britain': Space, Time and the Electric Telegraph in the Victorian Age," *British Journal for the History of Science* 33 (December 2000): 455–75.

15. Kirk Savage, *Standing Soldiers, Kneeling Slaves: Race, War, and Monument in Nineteenth-Century America* (Princeton: Princeton University Press, 1997).

16. Graeme Davison, *The Unforgiving Minute: How Australia Learned to Tell the Time* (Melbourne: Oxford University Press, 1993), 100. Davison explains that wireless, automobiles, and airplanes changed Australians' experiences of time and space.

17. About recent interest in public clocks, see Joe Heim, "For Public Clocks, a Time Warp," *Washington Post*, November 3, 2011. Heim reports that public clock manufacturers have annually sold two hundred to three hundred clocks since 2008, in contrast to a dozen annually in the late 1970s.

18. The phrase "the impulse to *wear* time" is from Mary Ann Doane, *The Emergence of Cinematic Time: Modernity, Contingency, the Archive* (Cambridge: Harvard University Press, 2002), 4, emphasis in the original.

19. Aveni, *Empires of Time*; E. G. Richards, *Mapping Time: The Calendar and Its History* (New York: Oxford University Press, 1998); Elisheva Carlebach, *Palaces of Time: Jewish Calendar and Culture in Early Modern Europe* (Cambridge: Belknap Press of Harvard University Press, 2011).

20. About the adoption of hours and days as standard measurements of time, see Dorhn-van Rossum, *History of the Hour*. About the transition from natural measures of time to the clock, see Sauter, "Clockwatchers and Stargazers," and Glennie and Thrift, *Shaping the Day*.

21. See O'Malley, *Keeping Watch*, 1–8, and Bartky, *Selling the True Time*, 7–12.

22. Bartky, *Selling the True Time*, 7–12.

23. Peter Galison adeptly considers the implications of time as a convention rather than a scientific truth in *Einstein's Clocks, Poincaré's Maps: Empires of Time* (New York: W. W. Norton, 2003). I adopt the reference to a time standard as a "gauge" from Clark Blaise, *Time Lord: Sir Sandford Fleming and the Creation of Standard Time* (New York: Pantheon Books, 2000), xiii. About the effort to bring the world under one time standard see Ian R. Bartky, *One Time Fits All: The Campaigns for Global Uniformity* (Stanford: Stanford University Press, 2007).

24. M. Norton Wise, introduction to *The Values of Precision*, ed. M. Norton Wise (Princeton: Princeton University Press, 1995), 7–9. Wise explains that "precision carries an immense weight" (3), which it did not prior to the nineteenth century. In his view, "precision has become the sine qua non of modernity" (11). Since the eighteenth century, scientists have used claims of precise and accurate timing to enhance the authority of science and scientific observations.

25. Sauter, "Clockwatchers and Stargazers."

26. Time Table of the Lowell Mills, 1851, 1853, and 1863, Timetables of Lowell Mills, 1851–87, Baker. The timetables were issued in October of each year setting the time standard for the following year, so the 1851 timetable was in effect for part of 1851 and most of 1852. The 1850s timetables identify the time standard as the "meridian time of Lowell," but then qualify this reference by stipulating which clock should be consulted in order to determine the meridian time.

27. Bartky, *Selling the True Time*.

28. F. Kelmo, *Kelmo's Watch-Repairer's Hand-Book* (Boston: A. Williams, 1869), 35.

29. Sauter, "Clockwatchers and Stargazers."

30. The Chicago Historical Society now owns *The Clock Mender*. It had been displayed on the seventh floor of Marshall Field's since 1948, when Rockwell gave it to the department store. The anecdote about Marshall Field's response to the painting was among the information that accompanied the painting when donated to the Chicago Historical Society. The eponymous "clock mender" might be interpreted as standing for the United States, who by implication was using an old-fashioned means (a pocket watch) to regulate the world (the clock) in the aftermath of the Second World War.

31. Sean O'Connor, "Businesspersons," in *Jazz Age People and Perspectives*, ed. Mitchell Newton-Matza and Peter Mancall (Santa Barbara: ABC-CLIO, 2009), 167–86, esp. 173.

32. Stuart Sherman, *Telling Time: Clocks, Diaries, and English Diurnal Form, 1660–1785* (Chicago: University of Chicago Press, 1996), 2, 8, 18, 36–37, 75, 117. Sherman explains: "The particular time proffered by the clocks, watches, and memorandum books so new and conspicuous in the period seemed to many serial autobiographers to limn a new temporality—of durations closely calibrated, newly and increasingly synchronized, and systematically numbered—durations that might serve as 'blanks' in which each person might inscribe a sequence of individual actions in an individual style" (18). Sherman disagrees with theorists like Walter Benjamin and others who argue that standard measures of time and space rendered each empty and homogenous. For an engaging interpretation of

the relationship between watches and diaries as timekeeping mechanisms, see also Rudolf Dekker, "Watches, Diary Writing, and the Search for Self-Knowledge in the Seventeenth Century," in *Making Knowledge in Early Modern Europe: Practices, Objects, and Texts, 1400–1800*, ed. Pamela H. Smith and Benjamin Schmidt (Chicago: University of Chicago Press, 2007), 127–42.

33. About increasing speeds of circulation as a fundamental characteristic of modernity, see Zygmunt Bauman, *Liquid Modernity* (Cambridge: Polity Press, 2000); James Gleick, *Faster: The Acceleration of Just about Everything* (New York: Pantheon, 1999); Dorthe Gert Simonsen, "Accelerating Modernity: Time-Space Compression in the Wake of the Aeroplane," *Journal of Transport History* 26 (September 2005): 98–117. Insightful explorations about the temporal conditions of modernity include Stephen Kern, *The Culture of Time and Space, 1880–1918* (Cambridge: Harvard University Press, 1983); Roger Friedland and Deidre Boden, "NowHere: An Introduction to Space, Time and Modernity," in *NowHere: Space, Time and Modernity*, ed. Roger Friedland and Deidre Boden (Berkeley: University of California Press, 1994), 1–60; Doane, *Emergence of Cinematic Time*; Joel Dinerstein, *Swinging the Machine: Modernity, Technology, and African American Culture between the World Wars* (Amherst: University of Massachusetts Press, 2003); A. Roger Ekirch, *At Day's Close: Night in Times Past* (New York: W. W. Norton, 2005). For a further discussion about the rising emphasis on the time kept by clocks and watches, consult Hassan and Purser, *24/7*.

34. With modernity arrived "a type of consciousness" that was, as literary critic Northrup Frye argues in *The Modern Century* (1967), "obsessed by a compulsion to keep up" (quoted in Malcolm Bradbury and James McFarlane, "The Name and Nature of Modernism," in *Modernism, 1890–1930*, ed. Malcolm Bradbury and James McFarlane [New York: Penguin Books, 1976], 22). About contingency, see Doane, *Emergence of Cinematic Time*. About scientific understandings of simultaneity, see Max Jammer, *Concepts of Simultaneity: From Antiquity to Einstein and Beyond* (Baltimore: Johns Hopkins University Press, 2006).

35. Kern, *Culture of Time and Space*; Adam Barrows, *The Cosmic Time of Empire: Modern Britain and World Literature* (Berkeley: University of California Press, 2011). Barrows argues, "The political dimensions of modernist temporality have been misrepresented in a long-standing critical tradition that equates modernist time with the private, interior and purely aesthetic pleasures of the Bergsonian *durée*" (8). He suggests instead that British modernists "sought to dislocate their own treatment of human temporality from its enlistment in the standard time system by resituating temporal processes within more meaningful, contextually determined and variable social patterns" (14–15). Moreover, Barrows asserts, "world standard time not only enabled the efficiency of advanced global imperialism, but more important (for a study of aesthetics), it provided English citizens with a conceptual tool for cognitively reading that new imperial space as intrinsically unified with England through the hyperprecision of Greenwich time" (17).

36. Galison, *Einstein's Clocks*; Jimena Canales, *A Tenth of a Second: A History* (Chicago: University of Chicago Press, 2010).

37. Doane, *Emergence of Cinematic Time*. In 2011, the collage-artist Christian Marclay's film *The Clock* reversed the cinematic compression and expansion of time through its real-time passage through the twenty-four hours of a day. Itself a timepiece, the film splices together moments from other films that visually or aurally portray an exact moment of clock time. See Roberta Smith, "As in Life, Timing Is Everything in the Movies," *NYT*, February 3, 2011; Daniel Zalewski, "The Hours," *New Yorker*, March 12, 2012, 51–63.

38. Civic authorities established master clocks in the following European cities: Geneva (1780), London (1792), Berlin (1810), Paris (1816), Vienna (1823). See Sauter, "Clock-watchers and Stargazers," 696.

39. E. P. Thompson, "Time, Work-Discipline, and Industrial Capitalism," *Past & Present* 38 (1967): 56–97. Dozens and dozens of scholars have used or revised Thompson's paradigm of time-discipline. The following list includes revisions that have been most useful for me: Thomas C. Smith, "Peasant Time and Factory Time in Japan," *Past & Present* 111 (1986): 165–97; Keletso Atkins, "'Kaffir Time': Preindustrial Temporal Concepts and Labour Discipline in Nineteenth-Century Colonial Natal," *Journal of African History* 29 (1988): 229–44; David Brody, "Time and Work during Early American Industrialism," *Labor History* 30 (1989): 5–46; Michael O'Malley, "Time, Work and Task Orientation: A Critique of American Historiography," *Time & Society* 1 (1992): 341–58; Akhil Gupta, "The Reincarnation of Souls and the Rebirth of Commodities: Representations of Time in 'East' and 'West,'" *Cultural Critique* 22 (Autumn 1992): 187–211; Mark M. Smith, "Time, Slavery and Plantation Capitalism in the Ante-Bellum American South," *Past & Present* 150 (February 1996): 142–68; Mark M. Smith, "Old South Time in Comparative Perspective," *AHR* 101 (December 1996): 1432–69; Kevin K. Birth, "Trinidadian Times: Temporal Dependency and Temporal Flexibility on the Margins of Industrial Capitalism," *Anthropological Quarterly* 69 (April 1996): 79–89; Paul Glennie and Nigel Thrift, "Reworking E. P. Thompson's 'Time, Work-Discipline and Industrial Capitalism,'" *Time & Society* 5 (1996): 275–99; Mark M. Smith, *Mastered by the Clock: Time, Slavery, and Freedom in the American South* (Chapel Hill: University of North Carolina Press, 1997); Henrik Agren, "Time and Communication: A Preindustrial Modernisation of the Awareness of Time," *Scandinavian Economic History Review* 49 (2001): 55–77; Sauter, "Clockwatchers and Stargazers." The full issue of *Japan Review* 14 (2002) explores the history of time discipline in early modern and modern Japan.

40. Sauter, "Clockwatchers and Stargazers," 709. Sauter observes at the end of his essay, "Only after people stopped disciplining clocks could clocks discipline people." Modern time discipline, he explains, "was not a product of the rise of the modern factory system, but was inextricably linked to changes in how early modern Europeans produced and distributed knowledge about the world."

41. About the figures responsible for introducing the ideals of accuracy and precision in the United States, see Carlene Stephens, *Inventing Standard Time* (Washington, DC: NMAH-SI, 1983); Carlene Stephens, "'The Most Reliable Time': William Bond, the New England Railroads, and Time Awareness in 19th-Century America," *Technology and Culture* 30 (1989): 1–24, esp. 7–8; Carlene Stephens, "The Impact of the Telegraph on Public Time in the United States, 1844–1893," *IEEE Technology and Society Magazine* 8 (March 1989): 4–10; Ian R. Bartky, "The Adoption of Standard Time," *Technology and Culture* 30 (1989): 25–56.

42. Sauter, "Clockwatchers and Stargazers," 693. Sauter asserts that "time discipline began as an urban product and emerged not from the factory floor but from the streets, where most people in the early modern world would have encountered clocks" (688).

43. Smith, "Old South Time." Smith explains that unlike rural populations elsewhere in the world, North Americans in rural areas "succumbed to clock time" because "mechanical timepieces, the railroads, and clock-regulated bells" supplanted the "commitment to natural time" (1445, 1446).

44. An exception to this oversight is Nick Yablon's *Untimely Ruins: An Archaeology of American Urban Modernity, 1819–1919* (Chicago: University of Chicago Press, 2009).

Histories of urban space that overlook public clocks include David Henkin, *City Reading: Written Words and Public Spaces in Antebellum New York* (New York: Columbia University Press, 1998); Mary P. Ryan "'A Laudable Pride in the Whole of Us': City Halls and Civic Materialism," *AHR* 105 (2000): 1131–70.

45. Barbara Adam, *Time* (Cambridge: Polity Press, 2004), 113; Smith, "Old South Time," 1452–53, 1462–66. Smith argues that because plantation slaves were subject to "clock-governed bells and horns" (1463), they exemplified time discipline. He characterizes Southern masters as "among the most effective, if ruthless, enforcers of clock time in the nineteenth century world" (1466). In acknowledging that "perhaps [plantation slaves] did not internalize time discipline," Smith leaves open the possibility that theirs was not a modern form of time discipline (1465). A modern form of time discipline is one in which the mandates of the clock are internalized, becoming, as Thompson puts it in his essay, "the inward notation of the time." Thompson also draws attention to this distinction by using the phrase "the inward apprehension of time" ("Time, Work-Discipline, and Industrial Capitalism," 57). Time obedience may be a more apt way to characterize the time regime under which slave labor operated than time discipline; each can serve capitalist ends within "modern" frameworks.

46. E. E. Evans-Prichard, *The Nuer* (New York: Oxford University Press, 1969), 103, quoted in Aveni, *Empires of Time*, 88.

47. In 1905, Elgin ordered a new tower clock for its factory building from Seth Thomas. It stipulated four 14′6″ sectional glass dials with black numerals and minute dots; the dials would be illuminated at night with an electrical attachment. Elgin's striking clock did not indicate the hours, but was programmed as follows: "78 blows at 6 a.m., 37 blows at 6:50 a.m., 1 blow at 7 a.m., 1 blow at 12 M, 37 blows at 12:50 p.m., 1 blow at 1 p.m., 1 blow at 5 p.m. on the first five working days of each week and to strike 1 blow at 4 p.m. on Saturday instead of at 5 p.m. and to remain silent on Sundays." Elgin paid $2,306 for the clock, which cost Seth Thomas $2,740 to make and install (book H, 122–24 [April 8, 1905], ST). The visibility of the clock in conjunction with the audibility of its time signals structuring the employee's days and weeks, exemplifies the complexity of modern time discipline, which here relies on both inward notations of the time leading to time discipline as well as facets of time obedience. The company town of Elgin ordered from E. Howard a tower clock for its city hall in 1893, with four 4′5″ dials and a 1,000-pound bell (box 9, 3:106, 122 [May 26 and July 15, 1893], HC).

48. Watch clocks trace their lineage to the watch that kept order over towns in ancient Rome and medieval Europe. For an overview of the earliest watch clocks in the United States, visit the Detex Watchman's Clock Album website, http://www.watchclocks.org/index.html (accessed May 12, 2011). The earliest notice about the watch clock I have found is an 1860 advertisement for John Polsey and Company, which promises "Improved Watch Clocks, for the detection of delinquent Watchmen while on duty." See *The Watch & Jewelry Trade of the United States: Containing a Full List of the Manufacturers of, and wholesale and retail dealers in watches, jewelry, etc. throughout the Union* (New York: William F. Bartlett, 1860), 15. The watch clock was E. Howard's most active line of business well into the 1910s. By 1904, E. Howard was keeping a separate account book for watch clock orders: one extant book covers the years between January 1904 and October 1918 (box 11, vol. 2, HC). My sample of the E. Howard account books covering some of the years between 1871 and 1905 included 315 orders for watch clocks.

49. In the 1880s, Willard Bundy and Alexander Dey innovated time-recording clocks, for which they each received patents. In the 1890s, D. M. Cooper in Rochester, New

York, developed a timecard system, and the Standard Time Stamp Company introduced automatic time stamps to verify when jobs were finished on the factory floor. About time clocks, see G. Russell Oechsle and Helen Boyce, *An Empire in Time: Clock & Clock Makers of Upstate New York* (Columbia, PA: NAWCC, 2003), 13–15, 37–38, 66–67; O'Malley, *Keeping Watch*, 160–61; Stephens, *On Time*, 161–65. About IBM and punch card systems in general, see Saul Englebourg, *International Business Machines: A Business History* (New York: Arno Press, 1976); Lars Heide, *Punch-Card Systems and the Early Information Explosion, 1880–1945* (Baltimore: Johns Hopkins University Press, 2009).

50. On the coexistence of multiple temporal orientations in the United States, see Thomas Allen, *A Republic in Time: Temporality and Social Imagination in Nineteenth-Century America* (Chapel Hill: University of North Carolina Press, 2008); Smith, "Old South Time" and *Mastered by the Clock*; Cheryl A. Wells, *Civil War Time: Temporality and Identity in America, 1861–1865* (Athens: University of Georgia Press, 2005); Aaron W. Marrs, "Railroads and Time Consciousness in the Antebellum South," *Enterprise & Society* 9 (September 2008): 433–57. On the layering of temporal regimes in communities across space and time, see Barbara Adam, *Timewatch: The Social Analysis of Time* (Cambridge: Polity Press, 1995); Greenhouse, *Moment's Notice*; Robert Levine, *A Geography of Time: The Temporal Misadventures of a Social Psychologist* (New York: Basic Books, 1997). Literary critic Thomas Allen makes the important point that clocks and watches are "loci for richly layered temporal experience," rather than sources of "a distinctly mechanistic form of temporality" (Allen, *Republic in Time*, 13).

51. Daniel Rodgers, *The Work Ethic in Industrial America, 1850–1920* (Chicago: University of Chicago Press, 1978); Alexis McCrossen, *Holy Day, Holiday: The American Sunday* (Ithaca: Cornell University Press, 2000).

CHAPTER ONE

1. The 1912 Meneely Bell Company catalog quotes William Sidney Gibson's essay "Church Bells" published in the *Quarterly Review* (1854). The catalog can be found in folder 34, box 66, ser. 2, Museum Records, 1873–1951, subser. A: Subject Files, IHA. Another use of the metaphor "the tongue of time" in relation to bells is Englishman William Harrison's *The Tongue of Time; or, The Language of a Church Clock* (1842; New York: Anson D. F. Randolph, 1853).

2. Alain Corbin, *Village Bells: Sound and Meaning in the 19th-Century French Countryside*, trans. Martin Thom (1994; reprint, New York: Columbia University Press, 1998), 112. See also Richard Cullen Rath, *How Early America Sounded* (Ithaca: Cornell University Press, 2003), 61–68; Frederick M. Shelley, *Early American Tower Clocks: Surviving American Tower Clocks from 1726 to 1870* (Columbia, PA: NAWCC, 1999), x–xiv; Mark M. Smith, *Listening to Nineteenth-Century America* (Chapel Hill: University of North Carolina Press, 2001), 35–39, 109–13, 273n11; Wu Hung, "Monumentality of Time: Giants Clocks, the Drum Tower, the Clock Tower," in *Monuments and Memory, Made and Unmade*, ed. Robert S. Nelson and Margaret Olin (Chicago: University of Chicago Press, 2003): 107–32.

3. About the distinction between time as a matter of measure and time as a matter of occasion, see Stuart Sherman, *Telling Time: Clocks, Diaries, and English Diurnal Form, 1660–1785* (Chicago: University of Chicago Press, 1996), 36–37, 75, 117.

4. David Cressy, *Bonfires & Bells: National Memory and the Protestant Calendar in*

Elizabethan and Stuart England (1989; reprint, Thrupp: Sutton Publishing, 2004), 16, 50–65, 67–87; Corbin, *Village Bells*, 111–14; Smith, *Listening to Nineteenth-Century America*, 10, 34–35, 52, 57–58, 87, 110–14.

5. Rath, *How Early America Sounded*, 47, 50, 66.

6. Most of the E. Howard and Seth Thomas orders for clockworks attached to Angelus strikers and chimes were from Catholic churches, but a few were placed by Congregational and Methodist churches. They all rang at 12 noon and 6 p.m., but the morning times that they were programmed to ring varied. St. Stephen's Church in New Orleans requested a morning chime for 5:25 a.m.; St. Raphael's Church on New York's working-class West Side requested a 6 a.m. strike; and in Bridgehampton, New York, the First Presbyterian Church wanted a 7 a.m. strike (book C, 149 [1887], book D, 49 [1888], and book C, 83 [1886], ST).

Most of the E. Howard and Seth Thomas orders for clockworks attached to Westminster chiming mechanisms were placed by wealthy Protestant churches in New York City and its environs, though several were in Boston, Chicago, and prosperous regional towns like Clinton, Illinois (box 9, 3:227 [1894], HC). St. Andrew's Church in Yonkers, New York, placed a representative order in 1889 for Westminster chimes to ring on four bells whose weight totaled three-and-a-half tons. It wished for the hour strike and chimes to be silent between 11 p.m. and 8 a.m. (book D, 76 [1889], ST).

7. E. P. Thompson, "Time, Work-Discipline, and Industrial Capitalism," *Past & Present* 38 (1967): 56–97, quotation 57.

8. On timekeeping in early modern Britain, see Robert Tittler, *Architecture and Power: The Town Hall and the English Urban Community, c. 1500–1640* (Oxford: Clarendon Press, 1991), 133, 138; also Paul Glennie and Nigel Thrift, *Shaping the Day: A History of Timekeeping in England and Wales, 1300–1800* (Oxford: Oxford University Press, 2009), 116–24. On the marking of time in Boston, Salem, and New York, see Smith, *Listening to Nineteenth-Century America*, 293n47; Chris DeSantis, *Clocks of New York: An Illustrated History* (Jefferson, NC: MacFarland, 2006), 74. For Philadelphia, see Frank M. Etting, *An Historical Account of the Old State House* (Boston: James R. Osgood, 1876), 27–29.

9. Rath, *How Early America Sounded*, 66; Shelley, *Early American Tower Clocks*, x–xiv.

10. Period source quoted in Harrold E. Gillingham, "New York City's First Town Clock," *NYHS Quarterly Bulletin* 20 (1936): 14.

11. Of the sixty churches built before 1800 in South Carolina, three had clocks and bells, four additional churches had only bells (Mark M. Smith, *Mastered by the Clock: Time Slavery, and Freedom in the American South* [Chapel Hill: University of North Carolina Press, 1997], 18–19).

12. Deborah Mathias Gough, *Christ Church, Philadelphia: The Nation's Church in a Changing City* (Philadelphia: University of Pennsylvania Press, 1995), 49; Benjamin Dorr, *A Historical Account of Christ Church, Philadelphia* (New York: Swords, Stanford, 1841), 18.

13. Quotation from Charles Peterson, "Early Architects of Independence Hall," *Journal of the Society of Architectural Historians* 11 (October 1952): 24. The first General Assembly met in the state house in 1735, but the building was not completed until 1741. In 1751, a tower with a steeple was added to the south side of the building. See Charlene Mires, *Independence Hall in American Memory* (Philadelphia: University of Pennsylvania Press, 2002), 1–8. According to two architectural historians, the steeple "soared over Philadelphia. It dominated the land for miles away; seen from boats in the river or carts on the road, from horseback or on foot, it was an arresting sight" (Henry Russell Hitchcock and

William Seale, *Temples of Democracy: The State Capitols of the USA* [New York: Harcourt Brace Jovanovich, 1976], 12–13).

14. About the Liberty Bell in the 1750s and 1760s, see Gary Nash, *The Liberty Bell* (New Haven: Yale University Press, 2010), 5–12; Independence Hall Association, "Liberty Bell Timeline," http://www.ushistory.org/libertybell/timeline.html (accessed May 18, 2011). About the state house clock, see James W. Gibbs, "America's Most Historic Clock: Its Fate and Descendants," *NB* 11 (December 1876): 306–12; Etting, *Historical Account of the Old State House*, 32; T. Mellon Rogers, "Independence Hall, Philadelphia, Pa., 1729–1896," addendum to diary dated December 1896, History Manuscript Collection, IHA; Willis P. Hazard, *Annals of Philadelphia and Pennsylvania in the Olden Time* (Philadelphia: J. M. Stoddart, 1879), 210. Very little is known about bells in the early republic.

15. Herbert B. Satcher, "Music of the Episcopal Church in the Eighteenth Century," *Historical Magazine of the Protestant Episcopal Church* 28 (1949): 377 (transcribed in "Bells" card file, drawers 2b and 3a, cabinet 1, IHA), dates the ring of bells as arriving in 1754. A few years later, Christ Church gave the newly built St. Peter's Church its old bells. About Christ Church's steeple, see the informative and scholarly article by Charles E. Peterson posted on the church's website: "The Building of Christ Church," http://www.christchurchphila.org/Historic-Christ-Church/Church/Scholarly-Articles/The-Building-of-Christ-Church/160/ (accessed March 13, 2012).

16. Smith, *Listening to Nineteenth-Century America*, 109. See also Mark M. Smith, "Old South Time in Comparative Perspective," *AHR* 101 (December 1996): 1432–69, esp. 1448–49.

17. Dava Sobel, *Longitude: The True Story of a Lone Genius Who Solved the Greatest Scientific Problem of His Time* (New York: Penguin Books, 1995). For a different spin on the story of John Harrison, the clockmaker who perfected the marine chronometer, see Glennie and Thrift, *Shaping the Day*, 279–406.

18. Silvio A. Bedini, "'That Awfull Stage' (The Search for the State House Yard Observatory)," in *Science and Society in Early America: Essays in Honor of Whitfield J. Bell, Jr.*, ed. Randolph Shipley Klein, Memoirs Series 166 (Philadelphia: American Philosophical Society, 1986), 155–99, see esp. 160–66, 199. About the observations of the transits of Venus and Mercury in 1769, see Andrea Wulf, *Chasing Venus: The Race to Measure the Heavens* (New York: Knopf, 2012).

19. Bedini, "State House Yard Observatory," 173, 195–97.

20. Etting, *Historical Account of the Old State House*, 32. Rittenhouse was paid £20 a year to regulate the state house clock. Etting writes, "Thus it was David Rittenhouse who regulated the clock, which prescribed the *time* to the Members of Congress in 1776" (32, emphasis in original). It is not clear why Duffield (1720–1801), a talented clockmaker, relinquished responsibility for the state house clock. About the state house clock between 1760 and 1778, see also Rogers, "Independence Hall." About clock making in eighteenth-century Philadelphia, see the essays in Frank L. Hohmann III, *Timeless: Masterpiece American Brass Dial Clocks* (New York: Hohmann Holdings, 2009). About Rittenhouse's achievements as a clockmaker, see Ronald R. Hoppes and Bruce Forman, *The Most Important Clock in America: The David Rittenhouse Astronomical Musical Clock at Drexel University* (Philadelphia: American Philosophical Society, 2009).

21. *Pennsylvania Archives*, 8th ser., 8, 6856, dated September 12, 1772, transcribed in Bells card file, IHA; Robert P. Reeder, "The First Homes of the Supreme Court of the

United States," *American Philosophical Society Proceedings* 76 (1936): 564–65, transcribed in Bells card file, IHA.

22. Alain Corbin describes several instances when complaints were made in rural France about the effect on the sick of the sound of bells (*Village Bells*, 300).

23. About timepiece ownership in the Middle Atlantic region, see Paul G. E. Clemens, "The Consumer Culture of the Middle Atlantic, 1760–1820," *William and Mary Quarterly* 62 (October 2005): 577–624, tables 9 and 10; Paul A. Shackel, *Personal Discipline and Material Culture: An Archaeology of Annapolis, Maryland, 1695–1870* (Knoxville: University of Tennessee Press, 1993), 96–99, 171, 180; Mark P. Leone, *The Archaeology of Liberty in an American Capital: Excavations in Annapolis* (Berkeley: University of California Press, 2005), 47, table 4; Richard Bushman, *The Refinement of America: Persons, Houses, Cities* (1992; reprint, New York: Vintage, 1993), 229.

About timepiece ownership in New England, see Paul B. Hensley, "Time, Work, and Social Context in New England," *New England Quarterly* 65 (1992): 531–59; Lee Soltow, "Watches and Clocks in Connecticut, 1800: A Symbol of Socioeconomic Status," *Connecticut Historical Society Bulletin* 45 (1980): 115–22; Alexis McCrossen, "The 'Very Delicate Construction' of Pocket Watches and Time Consciousness in the Nineteenth-Century United States," *Winterthur Portfolio* 44 (Spring 2010): 2–30.

About timepiece ownership in New York, see Charles F. Hummel, "An Economic, Social, and Art Historical Approach to Watch-Register Data, 1777–1827: An Early Report," presented to the Delaware Seminar (1995), typescript copy held in Winterthur Library; Martin Bruegel, *Farm, Shop, Landing: The Rise of a Market Society in the Hudson Valley, 1780–1860* (Durham: Duke University Press, 2002), table 9; Martin Bruegel, "'Time That Can Be Relied Upon': The Evolution of Time Consciousness in the Mid-Hudson Valley, 1790–1860," *Journal of Social History* 28 (1995): 547–64, esp. 551, table 1.

About timepiece ownership in the South, see Mark M. Smith, "Counting Clocks, Owning Time: Detailing and Interpreting Clock and Watch Ownership in the American South, 1739–1865," *Time & Society* 3 (1994): 321–39; Smith, *Mastered by the Clock*, 1, 34, and table A.25.

24. Leone, *Archaeology of Liberty*, 47, table 4. Leone does not differentiate between watches and clocks.

25. Henry Terry, *American Clock Making, Its Early History and Present Extent of the Business* (Waterbury: Henry Terry, 1870), 4. Terry's account is quoted in David Jaffee, *A New Nation of Goods: The Material Culture of Early America* (Philadelphia: University of Pennsylvania Press, 2010), 150, also 173. Jaffee shows that before Eli Terry and Simon Willard's innovations in the design and manufacture of clocks, a clockmaker could make twenty-five movements a year, selling them for twenty-five dollars each.

26. John Styles, *The Dress of the People: Everyday Fashion in Eighteenth-Century England* (New Haven: Yale University Press, 2007), 97–98; 323, table 12; 343, table 13; and 344, table 14. Styles's close study of watch thefts in eighteenth-century England reveals that few provincial Englishmen owned watches before the 1750s, but that by the 1770s this had changed. On levels of timepiece ownership generally, see citations in note 23 of this chapter.

27. Michael C. Harrold, *American Watchmaking: A Technical History of the American Watch Industry, 1850–1930*, supplement to *NB* (Columbia, PA: NAWCC, 1981), 1.

28. Notice reproduced in Richard Newman, "Colonial and Early American Watchmakers," *NB* 52 (December 2010): 694. The article reproduces many notices for watches from the period, as well as photos of some colonial-era watches.

29. Carter Harris, "Joshua Lockwood: An Early American Merchant and Craftsman," *NB* 45 (June 2003): 299–301. Lockwood did not sell clocks and watches during the Revolutionary War (300–301).

30. Styles, *Dress of the People*, 107. In a sample examined by Styles, no English watch owners had jobs in factories and only a few worked in transportation. It was while owners were at leisure that their watches were stolen (101–3).

31. Benjamin Franklin, *Autobiography and Other Writings*, ed. Kenneth Silverman (New York: Penguin Press, 1986), 31. John Styles recounts similar instances in eighteenth-century England in "Time Piece: Working Men and Watches," *History Today* 58 (January 2008): 44–50.

32. *Pennsylvania Packet* (Philadelphia), September 20, 1785. Thanks to Kirsten Sword for sharing this item with me.

33. The economic historian David Landes estimates that between 1775 and 1800, England produced 150,000–200,000 watches, of which less than half were exported. Despite the interruption of British-American trade during at least eight of those years (the war years) and limitations on American purchasing power through the 1780s and 1790s, it could be estimated that about 30,000 English watches arrived in the United States in the last quarter of the eighteenth century. See David Landes, *Revolution in Time: Clocks and the Making of the Modern World* (Cambridge: Belknap Press of Harvard University Press, 1983), 231.

34. "French Efforts at Colonial Trade," undated news clipping in vertical file, carton 4, HB5NGDF, MSS 598, Baker. For more notices advertising watches in colonial America, see J. Carter Harris, *The Clock and Watch Makers American Advertiser* (Great Britain: Antiquarian Horological Society, 2003).

35. Michael J. Sauter, "Clockwatchers and Stargazers: Time Discipline in Early Modern Berlin," *AHR* 112 (June 2007): 685–709, esp. 690–91.

36. The official website for the Liberty Bell suggests that the bell did not ring on the day of the Declaration of Independence, since the steeple was in bad condition. The rest of the bells of the city tolled throughout the day. See Independence Hall Association, "Liberty Bell Timeline." The historian Charlene Mires also presents this conclusion in *Independence Hall*, 20.

37. Nash, *Liberty Bell*, 40–48; Mires, *Independence Hall*, 80–82.

38. John Adams to Hezekiah Niles, February 13, 1818, in *The Works of John Adams*, ed. Charles Francis Adams (Boston, 1850–56), 10:283. "The colonies had grown up under circumstances of government so different there was so great a variety of religions, they were composed of so many different nations, their customs, manners, and habits had so little resemblance, and their intercourse had been so rare and their knowledge of each other so imperfect, that to unite them in same principles in theory and the same system of action was a very difficult enterprise. The complete accomplishment of it in so short a time and by such simple means was perhaps a singular example in the history of mankind. Thirteen clocks were made to strike together a perfection of mechanism which no artist had before effected."

39. Cressy, *Bonfires & Bells*; Corbin, *Village Bells*; Rath, *How Early America Sounded*.

40. Independence Hall Association, "Liberty Bell Timeline"; Nash, *Liberty Bell*, 18–20; Mires, *Independence Hall*, 24; Gibbs, "America's Most Historic Clock," 309.

41. Nash, *The Liberty Bell*, 20; David Kimball, *Venerable Relic: The Story of the Liberty Bell* (Philadelphia: Eastern National Park and Monument Association, 1989); Harold V. B. Boorhis and Ronald E. Heaton, *Loud and Clear: The Story of Our Liberty Bell* (Norristown, PA: Ronald E. Heaton, 1970); Mires, *Independence Hall*, 8, 26.

42. Joseph Leacock to James Trimble, June 30, 1800, transcribed in "Independence Hall Clock" files, drawer 4A, cabinet 3, IHA. The General Assembly still owned the building but was itself not subject to the irregularities of the clock, having moved out of the city. The capital of Pennsylvania moved to Harrisburg in 1812.

43. The market clock is visible in a nineteenth-century photograph and engravings of the Second Street Market (1745–1956): Prints Department, FLP; 1866 photograph, Prints Department, LCP; Hazard, *Annals of Philadelphia* (1879) 183. The High Street and Delaware Street clocks are discussed in "Report of the Committee on Public Clocks, etc.," March 12, 1835, Minutes of Special Committees of Council, 1834–35, reproduced in IH-CCF.

44. "Communication, State House Clock," *American Daily Advertiser* 28 (December 22, 1809): 3, transcribed in IH-CCF.

45. Bedini, "State House Yard Observatory," 177. About Rittenhouse's observatory, see Daniel K. Cassell, *A Genea-Biographical History of the Rittenhouse Family: And All Its Branches in America* (Philadelphia: Rittenhouse Memorial Association, 1893), 177.

46. About timepiece ownership, see citations in note 23 of this chapter.

47. Carlene Stephens, *On Time: How America Has Learned to Live by the Clock* (Boston: Bulfinch Press, 2002), 47–51.

48. Jaffee, *New Nation of Goods*, 170. Jaffee explains that "detailed instructions had to be provided for setting up the clocks, since they were sold by the Porters and their peddlers over a vast expanse of territory." After Terry successfully fulfilled the contract with the Porters, two of his employees, Silas Hoadley and Seth Thomas, bought his factory: it eventually was the basis of the clock company bearing the name Seth Thomas.

49. Ibid., 172–73; see also David Hounshell, *From the American System to Mass Production, 1800–1932: The Development of Manufacturing Technology in the United* States (Baltimore: Johns Hopkins University Press, 1984), 52–54.

50. Hounshell, *American System*, 52–54. About the antebellum clock industry in the United States, see also John Joseph Murphy, "Entrepreneurship in the Establishment of the American Clock Industry," *JEH* 26 (June 1966): 169–86; R. A. Church, "Nineteenth-Century Clock Technology in Britain, the United States, and Switzerland," *Economic History Review*, 2nd ser., 28 (November 1975): 616–30; David Landes, "Watchmaking: A Case Study in Enterprise and Change," *Business History Review* 53 (Spring 1979): 1–39; Donald Hoke, "British and American Horology: Time to Test Factor-Substitution Models," *JEH* 47 (June, 1987): 321–72; August C. Bolino, "British and American Horology: A Comment on Hoke," *JEH* 48 (September 1988): 665–67; Donald Hoke, "British and American Horology: A Reply to Bolino," *JEH* 49 (September 1989): 715–19.

51. See Thomas Allen's analysis of clocks displaying George Washington in *A Republic in Time: Temporality and Social Imagination in Nineteenth-Century America* (Chapel Hill: University of North Carolina Press, 2008), 75–78.

52. Michael O'Malley, *Keeping Watch: A History of American Time* (Washington, DC: Smithsonian Institution Press, 1996), 185; Bruegel, "'Time That Can Be Relied Upon,'" 553. Bruegel observes, "When clocks and watches became cheaper, their utility as time-keepers superseded their role as markers of social hierarchies" (554).

53. George Daniels, *English and American Watches* (New York: Abelard-Schuman, 1967), 47; Patricia A. Tomes, *Time at the Watch and Clock Museum* (Ephrata, PA: Science Press, 1988), 72; Chris Bailey, *Two Hundred Years of American Clocks and Watches* (Englewood Cliffs, NJ: Prentice Hall, 1975), 191–93.

54. Bruegel, "'Time That Can Be Relied Upon,'" 551, table 1; Smith, *Mastered by the*

Clock, 34, table 1. Smith notes that in 1815 Congress taxed imported clocks and watches, and that an 1842 tariff added clocks and watches to the ad valorem list. Taxes on timepieces were common (21, 23).

55. Because of the appearance of the names of more watch importers in antebellum city directories than in those from the second half of the eighteenth century, I argue that imports rose. The most significant increase was in the 1850s, but as early as 1820, the names of watch importers begin to appear in city directories. Figuring out how many watches were imported from France, England, and Switzerland is a puzzle. Most import and export statistics are expressed in terms of the value of the product, rather than in the number of units traded. So, for instance, France exported $59,404 worth of clocks in 1851, and $1,757,502 worth of watches in the same year (Smith, *Mastered by the Clock*, 24). What this means in terms of the number of clocks and watches exported to the United States is impossible to tell. Landes estimates that in 1800 European output of watches (mostly in England, France, and Switzerland) was between 350,000 and 400,000 pieces a year. By 1875, European output had increased to 2.5 million watches (Landes, *Revolution in Time*, 287). But again, determining how many of these watches arrived in the United States is nearly impossible.

56. Bedini, "State House Yard Observatory," 177. The archival holdings related to Riggs and Brother's date after 1887 and can be found in the following repositories: The J. Welles Henderson Archives and Library of the Independence Seaport Museum, the Winterthur Library, and the Hagley Museum and Library.

57. John H. Wilterding Jr. and Mike Harrold, "Early Industrial Watchcases," part 2, *NB* 47 (April 2005): 147–64, esp. 147 and 161.

58. Mires, *Independence Hall*, 38–46, 67; Max Page, "From 'Miserable Dens' to the 'Marble Monster': Historical Memory and the Design of Courthouses in Nineteenth-Century Philadelphia," *Pennsylvania Magazine of History and Biography* 119 (October 1995): 299–344, esp. 305.

59. Minutes of the Common Council, 5 (March 25, 1819–October 3, 1823), 128–29, transcribed in IH-CCF; Independence Hall Association, "Liberty Bell Timeline." On increased activity of the courts, see Page, "Courthouses," 306–8; on increasing emphasis on timing in business, see Bruegel, *Farm, Shop, Landing*.

60. Report of Clock Committee of Select and Common Councils, February 28, 1828, reprinted in "Proceedings of Councils" in *The Register of Pennsylvania* (Philadelphia, 1828), 1:152–54, emphasis in original. See also "Report of the Committee on Public Clocks" (1835).

61. Mires, *Independence Hall*, 67–73.

62. "Report of Clock Committee" (1828), 152–53. The conversion of the cost of the project from 1828 to 2010 dollars was made using the gross domestic product (GDP) deflator, which is the agreed-upon best measure for comparing the cost of construction projects over time. Calculation made using the information at Measuring Worth, http://www .measuringworth.com/uscompare/ (accessed March 28, 2011). For more details about the restoration of the Old State House, particularly its steeple, see Mires, *Independence Hall*, 73–77. John Wilbank cast the new bell.

63. Quotations from *Poulson's Advertiser*, February 9, 1828, transcribed in IH-CCF. See also "Report of Clock Committee" (1828), 154. The noted scientist Francis Gurney Smith suggested installing clock faces on the new steeple instead of the tower (*Poulson's Advertiser*, February 9, 1828).

64. *U.S. Gazette*, March 1, 1829, transcribed in IH-CCF.

65. *U.S. Gazette*, June 10, 1829, transcribed in IH-CCF; "Report of the Committee on Public Clocks" (1835). See also Gibbs, "America's Most Historic Clock," 309–10. St. Augustine's congregation paid for the upkeep and care of the hand-me-down clock and the ringing of the bell. Less than two decades later, the clock dials, bell, and movement were destroyed in the fires of the 1844 Nativist Riots.

66. Stephen Hasham (1823), quoted in Shelley, *Early American Tower Clocks*, ix. See also Shelley, *Early American Tower Clocks*, part 2; Chris DeSantis, *Clocks of New York: An Illustrated History* (Jefferson, NC: McFarland, 2006), 77–83; Donn Haven Lathrop's website, American Tower Clocks, 1717–2011, http://homepages.sover.net/~donnl/amtc .html (accessed February 7, 2012). In the 1950s, a sundial covered Faneuil Hall's clock's face; in the early 1990s, the clock was restored and the sundial removed. See Philip Bergen, "After Substantial Renovation, Old State House Reopens," *Alliance Letter: Monthly Newsletter of the Boston Preservation Alliance* 13 (August 1992): 4.

67. Mires, *Independence Hall*, 75; Independence Hall Association, "Liberty Bell Timeline."

68. About the trajectory of the Old State House's transformation into Independence Hall, see Mires, *Independence Hall*, 67–69, 99–113. About the conquest of time as part of Americans' sense of their destiny, see Allen, *A Republic in Time*.

CHAPTER TWO

1. George Lippard, *Quaker City; or, the Monks of Monk Hall* (1845; Philadelphia: George Lippard, 1847), 31. My thanks to Tom Allen for suggesting I look at *Quaker City* after he read a draft of this chapter.

2. Clayton Marsh, "Stealing Time: Poe's Confidence Men and the 'Rush of the Age,'" *American Literature* 77 (June 2005): 259–89, quotation 264–65.

3. David Henkin, *City Reading: Written Words and Public Spaces in Antebellum New York* (New York: Columbia University Press, 1998), 52, 24. Henkin does not include clocks in his study of city reading, even though some of them did include written words. Clock faces were read and scrutinized with as much care as the words found in public spaces.

4. Marsh, "Stealing Time," 259–65. See also Johannes Dietrich Bergmann, "The Original Confidence Man," *American Quarterly* 21 (1969): 561–77.

5. Edgar Allan Poe, "The Devil in the Belfry," in *The Works of the Late Edgar Allan Poe*, vol. 1, *Tales*, ed. Rufus W. Griswold (New York: J. S. Redfield, 1850), 383–91. The story was first published in 1839. I found the Edgar Allan Poe Society of Baltimore website helpful as I analyzed this story, http://www.eapoe.org/index.htm (accessed December 14, 2011). See also Marsh, "Stealing Time," 266–67.

6. The literary critic Raymond Williams developed the concept of "structures of feeling" to discuss widely shared perceptions and values among a generation. Raymond Williams, "Structures of Feeling" chap. 9 in *Marxism and Literature* (New York: Oxford University Press, 1977), 128–35.

7. "Report of the Committee on Public Clocks, etc.," March 12, 1835, Minutes of Special Committees of Council, 1834–35, reproduced in IH-CCF.

8. *The Laws of Etiquette; or, Short Rules and Reflections for Conduct in Society, by a Gentleman* (Philadelphia: Carey, Lea, and Blanchard, 1836), 81–82. I learned about this source from Marsh, "Stealing Time," 263. The entire passage reads: "If you make an ap-

pointment to meet anywhere, your body must be in a right line with the frame of the door at the instant the first stroke of the great clock sounds. If you are a moment later, your character is gone. It is useless to plead the evidence of your watch, or detention by a friend. You read your condemnation in the action of the old fellows who, with polite regard to your feelings, simultaneously pull out their vast chronometers, as you enter. The tardy man is worse off than the murderer. He may be pardoned by one person, (the Governor), the unpunctual is pardoned by no one."

Despite recommending punctuality, in another section, the guide advised against wearing a watch: "What has a fashionable man to do with time? Beside he never goes into those obscure parts of town where there are no public clocks and his servant will tell him when it is time to dress for dinner" (168–69). This comment I found discordant with the guide's other statements about timekeeping. Its reference to a servant suggests the passage was copied from a European etiquette guide, perhaps from the eighteenth century. Many nineteenth-century American etiquette guides were a pastiche of observations cribbed from various sources.

9. "Report of the Committee on Public Clocks" (1835).

10. Ibid. About Lukens's regulator and astronomical equipment, see "Catalogue of Town Clocks, Gold Chronometers, . . . Belonging to the Estate of the Late Isaiah Lukens, Dec'd. To be Sold at Auction" (Philadelphia, 1847), copy in the NAWCC Library.

11. "Report of the Committee on Public Clocks" (1835).

12. Frederick M. Shelley, *Early American Tower Clocks: Surviving American Tower Clocks from 1726 to 1870* (Columbia, PA: NAWCC, 1999), part 2; Donn Haven Lathrop's website, American Tower Clocks, 1717–2011, http://homepages.sover.net/~donnl/amtc.html (accessed February 7, 2012).

13. The Public Ledger Building (1842–67) can be seen in an undated photograph held by Prints Department, LCP, which also holds Conrad Kuchel's *Commissioners Hall, Spring Garden*, chromolithograph, ca. 1851.

14. Meneely Bell Company catalog (1906), in box 3, Warshaw-WC. Nearly all orders for striking clocks and bells from Seth Thomas and E. Howard specify the weight of the bells. In the 1870s and thereafter, bells weighing 1,000 pounds or more were regularly ordered. The following list is but a small sample from the Seth Thomas account books: St. Matthew's Lutheran Church (Hoboken, New Jersey) ordered a 1,000-pound bell in September 1877 (ST B 184); Brazos County Courthouse (Bryan, Texas) ordered a 1,015-pound bell in January 1878 (ST B 194); Peoria County Courthouse (Peoria, Illinois) ordered a 4,000 pound bell in March 1878 (ST B 198); the Ann Arbor City Hall ordered a 2,000-pound bell in April 1878 (ST B 208); St. Michael's Church (Beaver Dam, Wisconsin) ordered a 2,500-pound bell in January 1905 (ST H 110); the Yellowstone County Court House (Billings, Montana) ordered a 1,002-pound bell in March 1905 (ST H 114); the Libbey, Lindsay and Carr Company Store Building (Rochester, New York) ordered a 3,575-pound bell in June 1905 (ST H 139); Atlanta University ordered a 2,000-pound bell in July 1905 (ST H 144); and St. Michael's Church (New York City) ordered bells weighing 500, 800, 1,200, and 3,000 pounds in April 1911 (ST 1 261).

15. George McNeil, *The Labor Movement: The Problem of To-Day* (Boston: A. M. Bridgman, 1887), 345–46; "The Old Ship-Builders of New York," *Harper's New Monthly Magazine* 65 (July 1882): 223–41, esp. 233; "A Shipyard Poet Dies," *NYT*, March 9, 1907. During a strike for shorter hours during the 1870s, shipyard owners tried to destroy the bell; it was rescued and recast and, in 1880, was "rehung with much ceremony." In 1881, on its fiftieth anniversary, the children at the neighborhood grammar school were given

a half-holiday, while the shipyard workers convened to celebrate the bell's legacy ("After Fifty Years: First Ringing of the Eleventh Ward Bell after the Recasting of It," *NYT*, March 10, 1881). The bell changed locations several more times. In 1898, it ended up a relic in an area home for old shipbuilders where it stayed until 1940. See also David L. Perlman, "In Search of Mechanics' Bells," *Labor Heritage* 6 (Spring 1995): 64–71.

16. The literature on work time is extensive: see especially E. P. Thompson, "Time, Work-Discipline, and Industrial Capitalism," *Past & Present* 38 (1967): 56–97; Patrick Joyce, *Work, Society, and Politics: The Culture of the Factory in Later Victorian England* (New Brunswick, NJ: Rutgers University Press, 1980); David R. Roediger and Philip S. Foner, *Our Own Time: A History of American Labor and the Working Day* (New York: Greenwood Press, 1989); Hans-Joachim Voth, *Time and Work in England, 1750–1830* (New York: Oxford University Press, 2000).

17. McNeil, *Labor Movement*, 345.

18. Quotation from *PL*, February 26, 1846. On the removal of the Liberty Bell, see Independence Hall Association, "Liberty Bell Timeline," http://www.ushistory.org/libertybell/timeline.html (accessed May 18, 2011). On the bell's symbolism, see Mark M. Smith, *Listening to Nineteenth-Century America* (Chapel Hill: University of North Carolina Press, 2001), 177–81. Smith notes that the Liberty Bell activated meanings associated with "the ancient power of dismissing evil" and "the modern sound of democratic freedom" (181).

The abolitionist Darius Lyman sketched a fictional scene in which a slaveholder renounces both the Liberty Bell and the Declaration of Independence. Upon encountering the bell on a visit to Philadelphia, the slaveholder exhorts a Northerner: "Those words are rather fanatical. Liberty should not be proclaimed to all the land, to all the inhabitants thereof. That would disorganize society." The parable closes when "a voice proceeded from the bell," explaining that since its duty was to "speak to a nation of hypocrites," it suffered "this hideous crack" and could only issue forth a "miserable clatter" (Darius Lyman, "The Cracked Liberty-Bell," in *Leaven for Doughfaces; or, Threescore and Ten Parables Touching Slavery* [Cincinnati: Bangs, 1856], 114–16). Mark M. Smith mentions this parable in *Listening to Nineteenth-Century America*, 179.

19. Gary Nash, *The Liberty Bell* (New Haven: Yale University Press, 2010), 58–59, 136–37, 142, 147–48, 156–60, 178–79. The Liberty Bell's importance as a national emblem deepened over the course of the nineteenth and twentieth centuries, serving as a vehicle through which demands could be made to expand the domain of "liberty."

20. On repairs to the Old State House clock, see *PL*, April 4, 1845 (clock hands); *PL*, June 18, 1852, and July 1, 1853 (dials); "Telling Time by Electricity," *PL*, August 10, 1857 (fire alarm installation).

21. Capitalizing on twenty years of experimentation in the genre, the *Time Table*, initially published in 1865, was the first weekly publication devoted to railroad timetables. The first issue of the *Time Table* promised that it would be "delivered early Every Monday Morning." It is in the Broadside Collection of the Library of Congress (126, No. 2a). About railroads, clocks, and clock time in the antebellum era, see Aaron Marrs, *Railroads in the Old South: Pursuing Progress in a Slave Society* (Baltimore: Johns Hopkins University Press, 2009), 9, 87–97; Carlene Stephens, "'The Most Reliable Time': William Bond, the New England Railroads, and Time Awareness in 19th-Century America," *Technology and Culture* 30 (1989): 1–24; Ian R. Bartky, "The Invention of Railroad Time," *Railroad History* 148 (1983): 12–22; Mark M. Smith, *Mastered by the Clock: Time Slavery, and Freedom in the American South* (Chapel Hill: University of North Carolina Press, 1997), 79–92.

22. Quotation from Stephens, "The Most Reliable Time," 6–8, and see also 11–16; David Rooney and James Nye, "'Greenwich Observatory Time for the Public Benefit': Standard Time and Victorian Networks of Regulation," *British Journal for the History of Science* 42 (March 2009): 5–30.

23. Quotation from Ian R. Bartky, *Selling the True Time: Nineteenth-Century Timekeeping in America* (Stanford: Stanford University Press, 2000), 33, and see also 7–12, 19–31.

24. Elias Loomis, *The Recent Progress of Astronomy in the United States* (New York: Harper Brothers, 1856), 225; for a survey of astronomical observatories built between 1769 and 1853, see 202–92. Yale College's observatory opened in 1830, Harvard's in the 1840s, and in the 1850s, two privately owned observatories in and near New York City began operations.

25. Elias Loomis, "Astronomical Observatories in the United States," *Harper's New Monthly Magazine* 13 (June 1856): 25–52, quotation 51. When President John Quincy Adams urged Congress to erect an observatory, which he called "a light-house of the skies," the request was met with ridicule. Expressing its disdain for such measures, Congress made it clear in an 1832 act funding a survey of the coast that it would not authorize the construction of an observatory (Loomis, *Recent Progress of Astronomy in the United States*, 205–6).

26. Circular, Boston, July 1, 1859, in box 3, Business Records, 1808–1931, Bond Records; entries of March 31, 1852, October 26, 1853, and "List of Chronometers for Sale and on Loan, 1852," Daybook 1852–57, box 15, Bond Records. A catalog of Bond and Son's timepieces for sale in a public auction in 1876 included the following entry: "Lot 5: 45 Second-hand chronometers of various makers. These chronometers are at sea, loaned out" ("Catalogue of Chronometers, Watches, Clocks, etc.," folder 1, box 3, Business Records, Bond Records). The auction's first three lots included twenty-nine new chronometers, and the sixth lot included sixty-six "old chronometers, mostly not in order." About the rental and rating of chronometers, see Bartky, *Selling the True Time*, 50–74.

27. Stephens, "The Most Reliable Time," 11–16; Bartky, *Selling the True Time*, 73. See also the following articles, all by Carlene Stephens: "The Impact of the Telegraph on Public Time in the United States, 1844–1893," *IEEE Technology and Society Magazine* 8 (March 1989): 4–10; "Partners in Time: William Bond & Son of Boston and the Harvard College Observatory," *Harvard Library Bulletin* 35 (1987): 351–84; "Astronomy as Public Utility: The Bond Years at the Harvard College Observatory," *Journal for the History of Astronomy* 5 (1990): 21–35.

28. Bartky, *Selling the True Time*, 50–74.

29. Stephens, "The Most Reliable Time," 11. The verb "radiate" appears in relation to clocks in Galison, *Einstein's Clocks*, 46.

30. For an extended argument that time balls were largely symbolic, see Alexis McCrossen, "Time Balls: Marking Modern Times in Urban America, 1877–1922," *Material History Review* 52 (2000): 4–15. About the US Naval Observatory time ball, see Ian R. Bartky and Steven J. Dick, "The First North American Time Ball," *Journal for the History of Astronomy* 13 (1982): 50–54.

31. Caitlin Holmes, "The Astronomer, the Hydrographer and the Time Ball: Collaborations in Time Signaling, 1850–1910," *British Journal for the History of Science* 42 (September 2009): 381–406; Ian R. Bartky and Steven J. Dick, "The First Time Balls," *Journal for the History of Astronomy* 12 (1981): 155–64; P. S. Laurie, "The Greenwich Time Ball," *Observatory* 77 (1958): 113–15; and "The Time-Ball at Greenwich," *Nautical Magazine* 4 (1835): 584–86.

32. "Catalogue of Town Clocks, Gold Chronometers, . . . Belonging to the Estate of the Late Isaiah Lukens." The catalog lists the clock as costing $2,960. It also lists a "large Town Clock made for Louisville," that cost $2,000, and three large town clocks, two costing $900 and one $600. It claims that the "Clocks are *all new and perfect.*"

33. Under "Art and History," the US Senate website includes pages about the clocks found in the Senate. See in particular "Preserving Punctuality: The Old Supreme Court Clock," http://www.senate.gov/artandhistory/art/Art_Spotlights/Preserving_Punctuality .htm (accessed February 14, 2012).

34. Isaac Bassett, "A Senate Memoir," unpublished manuscript (1895) digitized on the US Senate website, http://www.senate.gov/artandhistory/art/special/Bassett/tdetail .cfm?id=11 (accessed February 14, 2012).

35. Bassett continued his recollection: "On one occasion when I was through turning the hands back Senator Hill of Georgia arose in his seat and called the attention of the Senate to the fact and declared that it was all wrong and there was nothing in the Constitution that warranted it. His remarks were winked at and several senators called out, 'All right Hill, Constitution or not,' which created a great laugh all over the Senate Chamber" (ibid.).

36. John Rodgers to Hon. John D. White, February 4, 1882, entry 4, 4:536–45, USNO-Records. In this lengthy letter to White, chairman of the House Sub-Committee of the Committee on Commerce, Admiral Rodgers, who at that time was superintendent of the USNO, explains in detail how time was determined and then suggests a plan for a "national prime meridian for purposes requiring continuous time and local time for the daily purposes of life." He proposes that Congress had the power to "compel all railroads which carry the mails to print continuous time based on Washington meridian in their time tables."

37. "A Standard of Time," *NYT*, February 16, 1856.

38. As one period source explains, "The Wall-street omnibuses were started in front of his store, and it was there that the phrase 'Benedict's Time' became the popular form of indicating correct City time" (obituary for Samuel Ward Benedict, *NYT*, May 5, 1880). See also advertisement for Benedict Brothers, *NYT*, December 14, 1886. In 1939, the American Telegraph and Telephone Company installed a precision clock in the window of their building, which stood on the site of Benedict's store (Donald Martin Reynolds, *Monuments and Masterpieces: Histories and Views of Public Sculpture in New York City* [New York: Thames and Hudson, 1988], 438).

39. "Time Ball Signals," *NYT*, March 8, 1856. See also, Bartky, *Selling the True Time*, 59–75.

40. "The Time-ball on the Custom-house," *NYT*, April 21, 1860; "Directions for Discovering When It Is Twelve O'Clock," *Vanity Fair*, April 28, 1860, 285; "The Time Ball," *Vanity Fair*, May 19, 1860, 334.

41. Paul G. E. Clemens, "The Consumer Culture of the Middle Atlantic, 1760–1820," *William and Mary Quarterly* 62 (October 2005): 577–624, tables 9 and 10; Smith, *Mastered by the Clock*, table A.25; Martin Bruegel, *Farm, Shop, Landing: The Rise of a Market Society in the Hudson Valley, 1780–1860* (Durham: Duke University Press, 2002), table 9; Bruegel, "'Time That Can Be Relied Upon': The Evolution of Time Consciousness in the Mid-Hudson Valley, 1790–1860," *Journal of Social History* 28 (1995): 547–64, esp. 551, table 1; Mark M. Smith, "Counting Clocks, Owning Time: Detailing and Interpreting Clock and Watch Ownership in the American South, 1739–1865," *Time & Society* 3 (1994): 321–39.

42. Benjamin E. Cook, Accounts, entries dated June 30, 1834, May 29, 1835, Decem-

ber 17, 1836, July 27, 1839, September 29, 1859, Gere Collection, Historic Northampton. Cook was a jeweler in Northampton, Massachusetts. The rental fees ranged from three to forty dollars. The rental period was not specified. About the rental of chronometers in port cities, see citations in note 26 of this chapter.

43. Quotation from undated flyer for the Jerome Manufacturing Company in box 2, Warshaw-WC. The same flyer describes how the operation cuts three billion feet of lumber to japan, veneer, or varnish for clock cabinets. About the Connecticut clock production levels, see David Hounshell, *From the American System to Mass Production, 1800–1932: The Development of Manufacturing Technology in the United States* (Baltimore: Johns Hopkins University Press, 1984), 57. About Yankee clock peddlers, see David Jaffee, *A New Nation of Goods: The Material Culture of Early America* (Philadelphia: University of Pennsylvania Press, 2010), 181–87; Joseph T. Rainer, "The Honorable Fraternity of Moving Merchants: Yankee Peddlers in the Old South, 1800–1860" (PhD diss., College of William and Mary, 2000), 93–97, 138–57; Rainer, "The 'Sharper' Image: Yankee Peddlers, Southern Consumers, and the Market Revolution," *Business and Economic History* 26 (Fall 1997): 27–43; Jaffee, "Peddlers of Progress and the Transformation of the Rural North, 1760–1860," *Journal of American History* 78 (1991): 511–35, esp. 517; Smith, *Mastered by the Clock*, 22–23. After the Civil War, peddlers no longer hawked clocks. About Black Forest clock peddlers in the US, see David R. Bailey, "Early Black Forest Clockmaking and Marketing," *NB* 49 (April 2007): 133–44, esp. 138. Since the 1760s, Black Forest clockmakers had been making small clocks without pendulums; their production reached fifteen million in the early nineteenth century.

44. None of the secondary literature on peddlers or salesmen mentions trade in watches or watchcases, but there is evidence that at least one peddler did sell watches, watch keys, and watch chains, see Abraham Burger Account Book, "Peddler's Record, 1849–1881," entries September 17, October 1 and 8, December 7, 1849, and December 18, 1850, Baker.

45. See Donald Hoke, *Ingenious Yankees: The Rise of the American System of Manufactures in the Private Sector* (New York: Columbia University Press, 1990), 186; Chris Bailey, *Two Hundred Years of American Clocks and Watches* (Englewood Cliffs, NJ: Prentice Hall, 1975), 194; Jaffee, *New Nation of Goods*, 187; and Hounshell, *American System*, 57–60.

46. About nineteenth-century international watch production and trade, see David S. Landes, *Revolution in Time: Clocks and the Making of the Modern World* (Cambridge: Belknap Press of Harvard University Press, 1983), 317–20. About the imports of watches during the antebellum period, see Michael C. Harrold, *American Watchmaking: A Technical History of the American Watch Industry, 1850–1930*, supplement to *NB* (Columbia, PA: NAWCC, 1981), 1. Harrold notes that the price of imported watches decreased between the 1820s and 1850s, but that even fifteen dollars for a Swiss watch would have placed it out of reach of most Americans, who earned no more than a dollar a day.

My assessment that watch imports rose in the 1850s is based on several qualitative factors. First, little about watches shows up in personal papers, newspaper accounts, or other reportage of daily life prior to the 1850s. Second, antebellum city directories list few jewelers, watch retailers, or wholesalers until the 1850s. By 1860, American watch importers had a separate listing in city directories: at New York's 81 Nassau Street, for instance, F. A. Kupfer represented Jules Rauss "importer of Watches from his own manufactory at Chaux-de-Fonds, Switzerland" (*Trow's New-York City Directory* [New York: Wilson's New-York Commercial Register, 1860], 29, held by NYHS). Through the Civil War years, more importers entered the trade, suggesting that both the business and the competition grew.

See the annual volumes of *Trow's New-York City Directory* held by the NYHS. By 1865, nearly a dozen firms in New York City claimed status as "importers of fine watches" (*Trow's New-York City Directory*, 101–2). Third, printed ephemera collections have many watch importers' receipts, broadsides, and posters from the 1850s, but only a handful from the 1840s, and none from earlier periods. See, for instance, the receipts and trade cards found in ser. 1, Scrapbooks, box 94, folder 1 (Jewelry), Landauer Collection.

47. Broadside, "S. Hammond & Co., Importers of Fine Watches," 1858, in box 2, Warshaw-WC. See also Hoke, *Ingenious Yankees*, 184–86, and John Reardon, *Patek Philippe in America: Marketing the World's Foremost Watch* (Maplewood, NJ: Cefari Publishing, 2008).

48. The flyer also read: "Our Watches are made of two solid pieces ONLY, one of the wood, and the entire pendant of Iron, making a *durable* Sign" ("Flyer for Emblematic Signs" [1855], in ser. 1, Scrapbooks, box 1, folder 2, Landauer Collection).

49. Wendy A. Woloson, *In Hock: Pawning in America from Independence through the Great Depression* (Chicago: University of Chicago Press, 2009), 105–9, quotations 106 and 107. Ulysses S. Grant pawned a watch for $22 two days before Christmas in 1857. The pawn ticket is extant because the pawnbroker's family saved it (109).

50. About demand for watches in Massachusetts, see Alexis McCrossen, "The 'Very Delicate Construction' of Pocket Watches and Time Consciousness in the Nineteenth-Century United States," *Winterthur Portfolio* 44 (Spring 2010): 1–30. About demand for watches in Long Island, see Charles F. Hummel, "An Economic, Social, and Art Historical Approach to Watch-Register Data, 1777–1827: An Early Report," presented to the Delaware Seminar (1995), typescript copy held in Winterthur Library. About demand for watches in antebellum New York City, see Richard B. Stott, *Workers in the Metropolis: Class, Ethnicity, and Youth in Antebellum New York* (Ithaca: Cornell University Press, 1990), 164, 174–75. About demand for watches in Connecticut, see Paul B. Hensley, "Time, Work, and Social Context in New England," *New England Quarterly* 65 (1992): 531–59; Lee Soltow, "Watches and Clocks in Connecticut, 1800: A Symbol of Socioeconomic Status," *Connecticut Historical Society Bulletin* 45 (1980): 115–22. About demand for watches in Delaware, see Clemens, "Consumer Culture of the Middle Atlantic," tables 9 and 10. About demand for watches in the Hudson Valley, see Bruegel, "'Time That Can Be Relied Upon'," 551, table 1; Bruegel, *Farm, Shop, Landing*, table 9. About demand for watches in the South, see Smith, *Mastered by the Clock*, table A.25; Smith, "Counting Clocks, Owning Time"; and Frederick F. Siegel, *The Roots of Southern Distinctiveness: Tobacco and Society in Danville, Virginia, 1780–1865* (Chapel Hill: University of North Carolina Press, 1987), 138–39.

51. For a delightful and insightful look at the fashion for wearing watches among workingmen in eighteenth-century Britain, see John Styles, "Fashioning Time: Watches," chap. 6 in *The Dress of the People: Everyday Fashion in Eighteenth-Century England* (New Haven: Yale University Press, 2007), 97–107; and Styles, "Time Piece: Working Men and Watches," *History Today* 58 (January 2008): 44–50. About how watches were worn in Europe, British colonies, and the United States, see Genevieve Cummins, *How the Watch Was Worn: A Fashion for 500 Years* (Woodbridge, UK: Antique Collectors' Club, 2010).

52. Styles, *Dress of the People*, 103.

53. On Lincoln's debate, see William Gienapp, *Abraham Lincoln and Civil War America* (New York: Oxford University Press, 2002), 31 (I thank Amy Greenberg for this reference). In 1863, Lincoln won a gold watch as a prize when he donated a handwritten version of the Emancipation Proclamation to a charity auction in Chicago. He wore the watch

attached to a thirteen-inch gold chain he received as a gift (Harold Holzer, "Presidential Gifts, circa 1860," *NYT*, February 12, 2001).

About the federal income tax levied between 1862 and 1872, see Cynthia G. Fox, "Income Tax Records of the Civil War Years," *Prologue* 18 (1986), available on the National Archives Records Association website, http://www.archives.gov/publications/prologue/1986/winter/civil-war-tax-records.html (accessed June 21, 2011).

54. The historian Mark M. Smith estimates that in 1860 five establishments in the South employed one or two watchmakers, while fifteen in the North employed an average of 8.6 people each. So it is probable that fewer than 150 skilled watchmakers worked in the United States (Smith, *Mastered by the Clock*, 26). In this context, "watchmaker" refers to a skilled artisan who could make a watch; there were also watch repairmen at work throughout the United States. About watchmakers at midcentury, see also Donald R. Hoke, *The Time Museum Historical Catalogue of American Pocket Watches* (Rockford, IL: Time Museum, 1991), 61–63, 102–5.

55. For more details about the early history of the Boston Watch Company (which would become the American Watch Company and then the Waltham Watch Company), see Harrold, *American Watchmaking*, 16–20; George Daniels, *English and American Watches* (New York: Abelard-Schuman, 1967), 47–48; Hoke, *Ingenious Yankees*, 184–96; Bailey, *American Clocks and Watches*, 198; Vincent P. Carosso, "The Waltham Watch Company: A Case History," *Bulletin of the Business Historical Society* 23 (December 1949): 165–87; Amy K. Glasmeier, *Manufacturing Time: Global Competition in the Watch Industry, 1795–2000* (New York: Guilford Press, 2000), 108–16; Hoke, *Time Museum Historical Catalogue*, 57–100; August C. Bolino, *The Watchmakers of Massachusetts* (Washington, DC: Kensington Historical Press, 1987).

56. Harrold, *American Watchmaking*, 19–21, 53. About the efforts to manufacture standardized watchcases in the 1850s, see John H. Wilterding Jr. and Mike Harrold, "Early Industrial Watchcases," part 2," *NB* 47 (April 2005): 147–64, esp. 147.

57. Harrold, *American Watchmaking*, 21.

58. Ibid., 21–22. Later in this book, I take up E. Howard's clocks. Harrold observes: "E. Howard never sacrificed quality to achieve quantity, so that the total production after more than 40 years was barely above 100,000 watches. His efforts established a reputation for building superior watches commanding superior prices" (22). About E. Howard's watches, see George Lewis Dyer, *The Story of Edward Howard and the First American Watch* (Boston: E. Howard Watch Works, 1910).

59. Harrold, *American Watchmaking*, 22, 25. Former Boston Watch Company employees formed the Nashua Watch Company in New Hampshire in 1859, with the aim of manufacturing a precision watch. Despite employing thirty-five people and accumulating enough material to make a thousand watches, by the spring of 1862 Robbins had acquired the bankrupt upstart company.

60. Wilterding and Harrold, "Early Industrial Watchcases," 147 and 149. Wilterding and Harrold emphasize the importance of Boss's patent, calling it "seminal" (149). They explain that owing to tariffs on watchcases, many European watches arrived in the United States uncased. As a consequence, "numerous case making shops in Eastern import centers" made "high-quality silver and gold cases in all styles." They were "frequently associated with watch importing firms as silent partners" (147).

61. Mark M. Smith makes this point near the end of *Mastered by the Clock*: "There was no new time consciousness simply because southerners had already developed an understanding of time both compatible with and complementary to the type of time essen-

tial to the new bourgeois social and economic order demanded by emancipation" (154). Cheryl A. Wells agrees that Civil War "did not permanently alter Americans' perceptions and usages of time." See Wells, *Civil War Time: Temporality and Identity in America, 1861–1865* (Athens: University of Georgia Press, 2005), 1.

62. Bruegel, "'Time That Can Be Relied Upon'"; Bruegel, *Farm, Shop, Landing*, 83–89. See also Christopher Clark, *The Roots of Rural Capitalism: Western Massachusetts, 1780–1860* (Ithaca: Cornell University Press, 1990); Jonathan Prude, *The Coming of Industrial Order: Town and Factory Life in Rural Massachusetts, 1810–1860* (New York: Cambridge University Press, 1983); Stuart Blumin, *The Emergence of the Middle Class: Social Experience in the American City, 1760–1900* (New York: Cambridge University Press, 1989).

63. Wells, *Civil War Time*, 7.

64. Ibid., 7–8. Wells explains that different clock times caused confusion among the troops and officers, which sometimes disrupted battle plans (9).

65. Henry C. Metzger to his father, January 12, 1865, Valley of the Shadow (accessed June 21, 2011).

66. Sylvester McElheney to Harriet McElheney, March 1, 1865, Valley of the Shadow (accessed June 21, 2011). Union sergeant Isaac Newton Parker went to great lengths to have his watch repaired (Wells, *Civil War Time*, 57). Confederate soldiers also repaired their watches (ibid., 57–58).

67. Franklin Rosenbery to John Rosenbery, November 1, 1863, February 18, 1864, October 16, 1864, Valley of the Shadow (accessed June 21, 2011).

68. George Thomas to Minerva Thomas, April 16, 1862, Thomas Family Correspondence, NDSC, http://www.rarebooks.nd.edu/digital/civil_war/letters/thomas/ (accessed June 21, 2011).

69. Wells, *Civil War Time*, 24–26; see also 15–22, 31.

70. Confederate lieutenant John Cheves Haskell, quoted by Wells (ibid., 34).

71. Ibid., 34–53. Wells uses Gettysburg to demonstrate that "battle time" dictated events on and off the battlefield during the first days and nights of July 1863. She also shows how other battles affected temporal regimes in hospitals and prison camps. However, it seems clear that it was the battle itself—not the temporal orientations, associated with "battle time"—that determined the events at Gettysburg, as well as in hospitals and prisons. About the framing of wartime events within religious temporal frameworks see Alexis McCrossen, *Holy Day, Holiday: The American Sunday* (Ithaca: Cornell University Press, 2000), 46–49.

CHAPTER THREE

1. David E. Hoxie appears in the 1868 Northampton city directory as an employee of D. F. Davison, who is listed as a jeweler with an office on Center Street (*Northampton Directory and Business Advertiser for 1868–69* [1868], held by Forbes Library, Northampton). Biographical information about Hoxie can be found in his obituary in the *Hampshire Gazette*, January 31, 1921, and in that paper's "Notable Men Series," published December 15, 1908. About a dozen letters Hoxie wrote to family members while in the army and shortly after he mustered out suggest that although he received a disability discharge, his injuries were minor. The letters are in a private collection.

2. In 1871, Hoxie is listed as a jeweler in business with Davison. The business is listed as "Davison & Hoxie, watches and jewelry, old P.O. building" (*Northampton Town Directory*

and Business Advertiser [1871], held by Forbes Library, Northampton). An informant for the R. G. Dun credit agency stated that Hoxie and Davison were "doing a small safe business worth two thousand" (R. G. Dun and Company, Credit Report Volumes, Massachusetts Series, vol. 46 [Hampshire County, vol. 1], entry for "Davison & Hoxie," April 1871, Baker). In 1873, Hoxie is listed as a jeweler, but no business is advertised and no address is given (*The Northampton, Holyoke, and Chicopee Directory, 1873–1874* [1873], held by Forbes Library, Northampton). The first notation in the city directory of a solo place of business for Hoxie is in the 1875 *Historical Register and General Directory of Northampton* (210), held by Forbes Library, Northampton.

3. I claim that there were many more watch repairmen than does Mark M. Smith, the only other historian to my knowledge who has considered this question. Relying only on US Census counts, Smith argues that the numbers of watch repairmen decreased between 1850 and 1870 (*Mastered by the Clock: Time Slavery, and Freedom in the American South* [Chapel Hill: University of North Carolina Press, 1997)], 24–29, tables A.7 and A.8). Census takers, following the custom of the day, did not differentiate between watch repair, watch assembly, and watchmaking; what is more, many jewelers engaged in watch repair and sales. Qualitative data extracted from sources such as city directories, newspaper advertisements, watch paper collections, and trade card collections suggest an increase in watch repairmen. About watchmakers at midcentury, see also Donald R. Hoke, *The Time Museum Historical Catalogue of American Pocket Watches* (Rockford, IL: Time Museum, 1991), 61–63, 102–5. For an interesting investigation into the history of watch repair in the United States, see the six-part series in *NB* 52 and 53 by the retired watch repairman David A. Christianson, in particular, "A Historical Census of Watchmakers," *NB* 52 (June 2010): 317–22; and "Watchmaking Education in America," *NB* 52 (August 2010): 418–26. See also Domenic J. Calzaretto, "Who Needs a Watchmaker?," *NB* 53 (August 2011): 418–20. Calzaretto suggests that there is a serious shortage of watch repairmen in the contemporary United States.

4. D. E. Hoxie, Daybook, Northampton, Massachusetts, 1876–84, Baker. Other extant watch registers include Dominy Family, Watch Registers, East Hampton, Long Island, New York, 1770–1852 (Winterthur Library and East Hampton Free Library); [anonymous watchmaker/repairman], Daybook, Newburyport, Massachusetts, 1802 (Winterthur Library); Edmund J. Schurmann, Daybook, Watch Repairs and Sales, Philadelphia, 1861–66 (HSP); P. A. Gardener, Watchmaker and Jeweler's Cashbook, 1887–90 (South Carolina Department of Archives and History); Fowler's Watch Shop, Repair and Sales Records, Boise, Idaho, 1898–1903 (Idaho State Historical Society).

5. Eighteen percent of all the watches Hoxie repaired were manufactured by the Waltham Watch Company; half of them were manufactured during the Civil War years. They accounted for more than 50 percent of all the American watches he repaired.

6. "The Watch as a Growth of Industry," *Appleton's Journal* 4 (July 9, 1870): 29–36, quotations 35–36. About the sentimental value placed on pocket watches, see Wendy Woloson, *In Hock: Pawning in America from Independence through the Great Depression* (Chicago: University of Chicago Press, 2009), 113–16.

7. "Watch as a Growth of Industry," 29–30. The article notes that "we wisely refused to parcel out the earth for the lineal glory of a few families" (30).

8. "Martha Washington's Watch," *NYT*, August 2, 1879. In 1940 the Mount Vernon Ladies Association, who had turned George and Martha Washington's home into a patriotic shrine, acquired the "gold watch used by Dr. James Craik when attending Gen. Washington during his last illness" ("Mount Vernon Ladies Receive Watch of Washington's Doctor," *Washington Post*, May 10, 1940).

9. Robin Pogrebin, "Timeless Lincoln Memento Is Revealed," *NYT*, March 11, 2009. Further engravings on the watch read "LE Grofs Sept 1864 Wash DC." The watch movement was English, but the gold case was American made. See also the 1906 *NYT* article in which the watch repairman, Jonathan Dillon, revealed the story about inscribing Lincoln's watch with his name and the date of the attack on Fort Sumter ("Who Has Lincoln's Watch?" *NYT*, April 20, 1906).

10. Amy K. Glasmeier, *Manufacturing Time: Global Competition in the Watch Industry, 1795–2000* (New York, Guilford Press, 2000), 131–35.

11. Hoxie repaired three "Centennial Watches," as recorded in his Daybook, entries 1,741 (1881), 2,116 (1882), and 2,374 (1883). On the 1870 Brooklyn watchcase, see "The Answer Box: Significance of Lady Elgin Watchcase Scene," *NB* 45 (October 2003): 623–24. On the "Betsy Ross" watches, see Andrew H. Dervan, "United States Watch Company (Waltham, MA) History and Watch Production," *NB* 49 (October 2007): 554–63, esp. 561. About the United States Watch Company's line of "President" watches, see Dervan, "United States Watch Company History," 559–61.

12. After Barack Obama's inauguration as president in 2009, the Presidential Watch Company was launched to sell the Jorg Gray 6500 Chronograph. See the company's website, http://barackswatch.com/ (accessed February 10, 2012).

13. Frequent references to exact clock time are found in the following letters and memoirs of soldiers who served in the Mexican-American War: Eba Anderson Lawton, ed., *An Artillery Officer in the Mexican War, 1846–7: Letters of Robert Anderson* (New York: G. P. Putnam's and Sons, 1911); George Ballentin, *The Mexican War, by an English Soldier* (New York: W. A. Townsend, 1860); Benjamin F. Scribner, *Camp Life of a Volunteer* (Philadelphia: Grigg, Elliot and Co., 1847); Emma Jerome Blackwood, ed., *To Mexico with Scott: Letters of Captain E. Kirby Smith to His Wife* (Cambridge: Harvard University Press, 1917). Thanks to Daniel Garza for sharing these references with me. George Ballentin describes the theft of a watch from an American sergeant who died on the field (56).

14. Cheryl A. Wells, *Civil War Time: Temporality and Identity in America, 1861–1865* (Athens: University of Georgia Press, 2005), 13–53, 58–62, 32. Wells shows that "each military tried to graft the antebellum authority of the clock" onto war activities (7).

15. In contrast to the argument being made here, Wells suggests that "battle time reordered camp life" away from the clock, "by forcing soldiers to revert to the premodern task system" (ibid., 69). I agree with Michael O'Malley that the task system was not necessarily "premodern," and that it could coexist with time orientation. See O'Malley, "Time, Work and Task Orientation: A Critique of American Historiography," *Time & Society* 1 (1992): 341–58.

16. Watch pocket in general's jacket: item numbered Z.3083; watch pockets in soldiers' trousers: items numbered 1947.504b, 1950.3b, and Z.3091, Luce Center. See also Wells, *Civil War Time*, 57.

17. The gold watch with a shield engraved on its case was presented to General R. C. Rutherford by "his friends and fellow officers" in December 1861. Twenty men who had been under his command in Cairo, Illinois, donated between $2 and $10 each to purchase the $102 English gold pocket watch (item number 1942.439, Luce Center). See also "Handwritten Testimonial, Cairo, Ill., December 1861," Miscellaneous Manuscripts—R. C. Rutherford, NYHS Manuscript Collections.

18. D. G. Farragut's pocket watch, held by NAWCC Museum, dates from around 1859, maker unknown. Farragut was an admiral in the Union navy. A watch had long been considered a necessity for maritime navigation. Indeed, a gold watch recovered from the

Confederate submarine *H. L. Hunley*, which sunk to the bottom of the Charleston harbor February 17, 1864, after sinking an enemy ship, belonged to the ship's commander, Lieutenant George Dixon. See Brian Handwerk, "Watch May Help Solve Mystery of Civil War Sub *Hunley*," http://news.nationalgeographic.com/news/2003/03/0310_030310_Hunley.html (accessed March 10, 2003).

19. Item number 1953.195t.1-3, Luce Center.

20. Carlene Stephens, *On Time: How America Has Learned to Live by the Clock* (Boston: Bulfinch Press, 2002), 95.

21. John M. Jackson to Alonzo Jackson, September 30, 1862, John M. Jackson Letters, NDSC. A few months later, Jackson described how he put a watch pocket in his clothes: "I feel quite pleased with how it looks" (John Jackson to Joseph Jackson Family, December 4-5, 1862). Charles H. Brewster, of Northampton, Massachusetts, sent his watch to his mother for repairs in January 1863, but in a letter to her sent a month later he writes: "You need not do anything about the watch as I have got one here and poor at that, yet private soldiers buy them by [the] dollars worth" (Charles H. Brewster to Martha C. Brewster, February 26, 1863, in Civil War Letters of Charles Harvey Brewster Collection, Historic Northampton, A.C.W. 18.269–A.C.W. 18.287, folder 8).

22. A lieutenant in the Union army asked his brother to buy "a lady's gold Hunting watch for a Christmas present to Fannie; get a small one and as nice as you can, and give it to her on Christmas day. Have it marked with her name in full. Send me the price and I will forward all money" (William Latham Candler to John C. Candler, December 8, 1861, William Latham Candler Papers, 1861-63, Ms 97-007, Virginia Tech Digital Library and Archives). The lieutenant who commanded the Confederacy's submarine *Hunley* also gave his fiancée a gold watch on Christmas day 1862. Brian Hicks, "Clue to Dixon's Identity Locked in Queenie's Locket," *Post and Courier*, March 2, 2003, www.charleston.net (accessed June 21, 2011).

23. The gift of watch and revolver of the Ninth Pennsylvania is recorded in a letter from Bob Taggart to Captain John Taggart, July 26, 1861, Valley of the Shadow. Many personal letters discuss gifts like watches and guns.

24. A small Hubbard Brothers catalog is extant: "Catalogue of Watches and Gold Pens and Holders of the Manufacture and Importation of the Hubbard Brothers, No. 169 Broadway, New-York," ca. 1865, ser. 1, Scrapbooks, box 94, folder 1 (Jewelry), Landauer Collection.

25. Luman Ballou, War Journal, 1864 entries dated April 16, April 21, April 26, April 27, May 8, May 12, and May 15. Ballou served in the Seventh Infantry, Vermont Volunteers. The journal is digitized on the website VermontCivilWar.org, http://vermontcivilwar.org/units/7/64a.php (accessed June 21, 2011).

26. Rufus Ricksecker to "Folks at Home," April 9 and 10, 1863. Ricksecker, a member of the 126th Ohio Volunteer Infantry stationed in Martinsburg, Virginia, also displayed a keen awareness of time and its passage. In a letter dated May 19, 1863, he wrote: "There are only about 41 men in camp, so you can imagine how time hangs & how lonesome & desolate camp looks & is." The Rare Books Library of the Ohio State University, Columbus, Ohio, holds Ricksecker's original letters, http://www.frontierfamilies.net/family/Rickpt1.htm (accessed June 21, 2011).

27. William H. Johnson to his family, May 19, 1862, quoted on the William H. Johnson Civil War Letters page from the website of the University of South Carolina, South Caroliniana Library, http://library.sc.edu/socar/uscs/1995/whjohn95.html (accessed June 21, 2011). Johnson, a sergeant in Company E, Seventh Connecticut Volunteers, was stationed in South Carolina and Georgia between 1861 and 1863. About trading watches

while enlisted, see also Union army private Harrison E. Randall Letters, NDSC, http://www.rarebooks.nd.edu/digital/civil_war/letters/randall/ (accessed June 21, 2011).

28. Rachel Cormany, diary entry, June 28, 1863, Valley of the Shadow (accessed June 21, 2011).

29. Joseph Addison Waddell, diary entry, May 4, 1865, Valley of the Shadow (accessed June 21, 2011). The entry reads: "Some of the returned negroes say the Yankee soldiers robbed them of money, watches, and all valuables." Waddell lived in Staunton, Virginia, and was a Confederate.

30. T. H. Mann, "A Yankee in Andersonville," part 1, *Century Magazine* 40 (July 1890): 449; A. S. Abrams, *The Trials of the Soldier's Wife: A Tale of the Second American Revolution* (Atlanta: Intelligencer Steam Power Presses, 1864), 37–42; Wells, *Civil War Time*, 93. A letter from one prisoner to his wife instructed, "If you have not sent my watch do not send it" (Stephen F. Sullivan to Lucy J. Sullivan, May 3, 1864, Andersonville/Wirz Collection, NDSC, http://www.rarebooks.nd.edu/digital/civil_war/topical_collections/andersonville/index.shtml [accessed June 21, 2011]).

31. Photograph of black soldier seated with pistol in hand, watch chain in pocket, ninth-plate ambrotype, ca. 1860–70, LC-USZ62-132207, LC-PPD.

32. Mechanical timepiece ownership increased from 50 percent to 69 percent among members of the lowest wealth quintile in Charleston and from 40 percent to 68 percent among the same quintile in Laurens County, South Carolina (Smith, *Mastered by the Clock*, 34, table 1, and tables A.15–A.16, A.21–A.22).

33. Incidentally, this customer sold or traded the watch sometime during the five years after he had it repaired, since another customer brought it in for repairs six years later (Hoxie Daybook, entry 2220). Most Northampton volunteers for the Union army joined the four infantry regiments raised in Western Massachusetts. Hoxie joined Company C, which was chartered in 1801 as the city's militia company, of the Tenth Massachusetts Regiment. On July 25, 1861, this unit—with 1,000 men, 50 from Northampton—left Boston's harbor. In its 1864 homecoming parade in Springfield, Massachusetts, only 220 men marched. Hoxie suffered injuries in the field that left him paralyzed, which is what precipitated his discharge.

34. Joseph Hugh Green, *Jewelers to Philadelphia and the World: 125 Years on Chestnut Street* (New York: Newcommen Society in North America, 1965), 8–12.

35. Saul Engelbourg, "The Economic Impact of the Civil War on Manufacturing Enterprise," *Business History* 21 (July 1979): 148–62, quotation 157. About the effect of the Civil War on the American economy as a whole, see Stanley Engerman and J. Matthew Gallman, "The Civil War Economy: A Modern View," in *On the Road to Total War: The American Civil War and the German Wars on Unification, 1861–1871*, ed. Stig Förster and Jörg Nagler (Washington, DC: German Historical Institute / New York: Cambridge University Press, 1997), 217–48; and Phillip Shaw Paluden, "What Did the Winners Win? The Social and Economic History of the North during the Civil War," in *Writing the Civil War: The Quest to Understand*, ed. James McPherson and William Cooper Jr. (Columbia: University of South Carolina Press, 1998): 174–200.

36. Treasurer's Reports for 1860, 1861, 1862, Waltham Records. Between 1860 and 1861, the company sold 14,351 watch movements. In 1862, it sold 20,341 movements and 7,700 cases; at the end of that year, it had 3,278 movements and 1,959 cases in stock.

37. Treasurer's Report for 1863, Waltham Records. The company sold 39,188 movements and 11,506 cases; it had 2,167 movements and 3,253 cases in stock.

38. Speech by M. Edouard Favre-Perret, November 14, 1876, reprinted in the appendix to James C. Watson's *American Watches: An Extract from the Report on Horology at the International Exhibition at Philadelphia, 1876* (New York: Robbins and Appleton, 1877), 33.

39. Treasurer's Report for 1864, Waltham Records. The company sold 36,376 movements and 11,719 cases; it had 10,373 movements and 6,604 cases in stock. About the five watch manufacturers who began production in 1864, see Michael C. Harrold, *American Watchmaking: A Technical History of the American Watch Industry, 1850–1930*, supplement to *NB* (Columbia, PA: NAWCC, 1981), 27–34.

40. Ibid., 27–28, 30, 32–33.

41. Ibid., 28. Aaron Dennison, who had been instrumental in founding what became the American Watch Company, organized the Tremont Watch Company along what we today would call "off-shoring principles." He set up a workshop in Zurich, where skilled labor was cheaper than in the United States, to make the intricate watch parts (trains, escapements, and balances), while a Boston workshop made the flat steel parts, plates, and barrels. This setup worked, but the owners decided to move all production back to Boston, renaming the company the Melrose Watch Company; by the end of 1868, it had failed (ibid., 32).

42. Ibid., 28–34. The Waltham Watch Company's treasurer listed the names of his company's domestic competitors (Treasurer's Report for 1885, Waltham Records.)

43. See John H. Wilterding Jr. and Mike Harrold, "Early Industrial Watchcases," part 2, *NB* 47 (April 2005): 147–64, and Warren H. Niebling, *History of the American Watch Case* (Philadelphia: Whitmore, 1971). In 1885 John Dueber purchased a majority share of the Hampden Watch Company, which he moved close to his Ohio and Kentucky watchcase factories. See Michael C. Harrold, "Charles Rood and Henry Cain: Origins of the Hamilton Watch Company," *NB* 44 (October 2002): 553–54.

44. For archival collections of trade cards related to pocket watches, see boxes 1–3, Warshaw-WC; Trade Card Collection, NAWCC Library; ser. 1, Scrapbooks, box 94, folder 1 (Jewelry), Landauer Collection. For archival collections of trade catalogs related to pocket watches, see Hagley Museum and Library; NAWCC Library; Archives Center, NMAH; NMAH-SI Library; NMAH-SI Division of Technology.

45. Wilterding and Harrold, "Early Industrial Watchcases"; Howard Lasser, "The Story of Hayden W. Wheeler," *NB* 47 (October 2005): 548–60. In the early 1870s, Hayden Wheeler (1827–1904), a jeweler working in New York City, opened the Brooklyn Watch Case Company, which used machines to manufacture watchcases. Wheeler's son, Willard Hayden Wheeler (1863–1923) began collecting watches in 1880; his world-famous collection was displayed at the Brooklyn Museum for years, before being sold at auction by Sotheby's in October 1961.

46. Alexis McCrossen, "The 'Very Delicate Construction' of Pocket Watches and Time Consciousness in the Nineteenth-Century United States," *Winterthur Portfolio* 44 (Spring 2010): 1–30, esp. 10, 12, table 3.

47. Favre-Perret speech (1876), 33. For an interesting look at the Swiss response to the 1876 exhibit of American-made watches, see Richard Watkins, "Jacques David—and a Summary of 'American and Swiss Watch-making in 1876' with Emphasis on Interchangeability in Manufacturing," *NB* 46 (June 2004): 294–302. This article provides details about Jacques David's report about American watchmaking based on his 1876 visit to the Philadelphia Centennial and to several watch factories. According to Watkins, the Swiss ignored the 108-page report, prompting David to write a shorter 12-page version in 1877.

48. Table 5 in David S. Landes, *Revolution in Time: Clocks and the Making of the Modern World* (Cambridge: Belknap Press of Harvard University Press, 1983), estimates the value of Swiss watch exports to the United States (1864–82), but the data is unreliable and difficult to translate into the numbers of watches imported in any given year. A Swiss watchmaker, Georges-Frédéric Roskopf, introduced the first "cheap watch," which came to be known as the "proletarian watch." See Eugène Jaquet and Alfred Chapuis, *The Swiss Watch* (Switzerland, 1953), 187–88, plate 123; Glasmeier, *Manufacturing Time*, 121–24. Two of Hoxie's customers had Roskopf watches, one brought his in three times for cleaning and repairs (Hoxie, Daybook, entries 758, 999, 1504, 1771).

49. David Landes, "Watchmaking: A Case Study in Enterprise and Change," *Business History Review* 53 (Spring 1979): 1–39, esp. 18–20 and 31–34.

50. McCrossen, "Pocket Watches and Time Consciousness," 10, table 3. A catalog issued in the mid-1870s by New York importers Cross and Beguelin (held in Landauer Collection) offers what likely are Swiss-made imitation American watches under the signatures "Centennials" (offered for between $8 and $12.50) and "Democratic" (for $5.50 to $5.75). About imitation American watches, see Landes, *Revolution in Time*, 324–25. About pawning fake watches during the nineteenth century, see Woloson, *In Hock*, 142–44.

51. Watches made by the Waltham Watch Company won first place in the horology competitions held under the auspices of the 1876 and 1885 world's fairs. See Watson, *American Watches*, 3–30, 39–40; American Watch Company, *1884–1885 New Orleans Exposition Catalog* (1884; reprint, Fitzwilliam, NH: Ken Roberts, 1972).

52. See the Treasurer's Report for 1876, Waltham Records, and Favre-Perret speech (1876), 31–39. The Waltham treasurer explained that the company "entered twelve movements against [the Swiss]." After six months of testing, four of their movements were judged the best in the competition. The success increased foreign demand for Waltham watches.

53. Glasmeier, *Manufacturing Time*, 122.

54. Folder 2, "Advertisements: Other Watch Companies," box 3, William Bond and Son Business Records, 1808–1931, Bond Records.

55. About the Jewelers Mercantile Agency, see "Bon Voyage to the United Associations," *Jewelers' Circular* 38 (June 28, 1899): 16–19. The Hagley Museum and Library holds the 1884 edition of the Jewelers Mercantile Agency's *Directory of Wholesale Dealers (manufacture & jobbers) of jewelry, watches, clocks, diamonds & precious stones—in the United States*. An earlier type of directory for the trade, issued in 1860 and 1861, included manufacturers, wholesalers, and retailers in watches and jewelry. The Library Company of Philadelphia holds *The Watch & Jewelry Trade of the United States: Containing a Full List of the Manufacturers of, and wholesale and retail dealers in watches, jewelry, etc. throughout the Union* (New York: William F. Bartlett, 1860) and a slightly different edition published in 1861.

56. Jewelers Mercantile Agency, *Directory of Wholesale Dealers* (1884), 81, 94–97, 100, 116, 170–81. In addition to the watch movement dealers, it lists sixteen watch case manufacturers and wholesalers (181–82), two wholesalers of watch glasses (184), two of watch keys (184), and four of watchmakers' regulators (184). About New York City's Nassau Street and Maiden Lane, see Lasser, "Story of Hayden W. Wheeler," 549.

57. Michael C. Harrold, "An Economic Look at the American Watch Industry: The Pocket Watch Era, 1860–1930," *NB* 49 (August 2007): 425–41, esp. 426.

58. William Dunn, "The Waterbury Rotary Watch," *NB* 52 (October 2010): 542–53.

Benedict and Burham Manufacturing Company first started making brass clocks in 1857 in Waterbury, Connecticut. In 1878, it got into the watch business, naming the enterprise the Waterbury Watch Company. In 1898, the Waterbury Watch Company was reorganized as the New England Watch Company. In 1914, Ingersoll bought it. But in 1922, the Waterbury Clock Company bought Ingersoll, renaming it the "Ingersoll Division of Waterbury Clock Company." In 1944, the division was sold to US Time Corporation, which was renamed Timex in 1969.

59. See Michael Harrold, "Manhattan Watch Company with Notes on the Knickerbocker Watch Co.," part 1, *NB* 52 (April 2010): 145–59, and Harrold, "Charles Rood and Henry Cain," 555–56. The Keystone Watch Club Company was renamed the Lancaster Watch Company in 1889.

60. *Independent Watch Company* (Fredonia, New York), 1877, box 2, Warshaw-WC; advertisement for St. Louis's Mermod and Jaccard Jewelry Company, *Bates County Record* (Butler, MO), November 17, 1883; Joseph Gardner and Son, *Illustrated Catalogue of Watches and Description of Ball Clock in Their Window* (Boston, [1879]), oversized, folder 6, box 203, Warshaw-WC; advertising broadside, "Do You Want a Watch this Fall?" (Mineral Wells, Texas, [ca. 1895–1905]), DL-SMU. My thanks to Russell Martin, director of the DeGoyler Library, for sharing this last item with me.

61. See Richard S. Tedlow, *New and Improved: The Story of Mass Marketing in America* (Boston: Harvard Business School Press, 1996), 263–65. Tedlow estimates that Sears sold the business for "at least $60,000 and perhaps as high as $100,000" (263). Montgomery Ward's *Catalogue No. 57* (1895) advertises many more types of watches than clocks; compare listings for clocks (198–202), men's watches (135–43), boy's watches (144), ladies watches (145–50), watchmakers' tools (151), and even toy watches for boys and girls (229).

62. See Landes, *Revolution in Time*, 317–20, tables 3 and 4, detailing the output of the Waltham Watch Company (1857–1953) and the Elgin Watch Company (1867–1953). Independent historian and pocket-watch expert Michael C. Harrold's numbers differ from the output statistics Landes compiled. See Michael C. Harrold, "Charles Rood and Henry Cain" and "Economic Look at the American Watch Industry." Although the extent of the sources Harrold used for his data is unspecified, it may be that his numbers are the more accurate of the two sets.

63. Harrold, "Charles Rood and Henry Cain," 557. Harrold describes how Dueber was able to work around the "watch trust," officially known as the "National Association of Jobbers in American Movements and Cases." Waltham, Elgin, Hampden, and the Illinois Watch Company supplied jobbers who annually made $5,000 worth of purchases from the association and who sold watches, cases, and parts at 25 percent or more above the wholesale price. Retailers' only access to these leading brands of American-made watches was through these jobbers. Dueber found the terms restrictive and began to sell his watches and cases directly to retailers.

64. Treasurer's Report for 1866, Waltham Records. Waltham's 1874 report describes efforts to secure foreign markets: Waltham set up agency in London and initiated trade with "the leading States of South America, the West-Indies, Great Britain, and her principal colonies, China, Japan, and some of the countries of Europe." The treasurer mentions that gold watches were bought in New York for export to Japan (Treasurer's Report for 1874, Waltham Records.)

65. Members of the band also worked in the Elgin factory but were given time off to play in venues such as the Ice Palace in St. Paul, Minnesota; the Corn Palace in Mitchell,

South Dakota; and the Republican National Conventions of 1888 and 1892, in Chicago and Minneapolis respectively. See E. C. Alft, *Elgin: An American History, 1835–1985* (Elgin, IL, 1984), 60, 63.

66. Michael O'Malley, *Keeping Watch: A History of American Time* (Washington, DC: Smithsonian Institution Press, 1990), 174–79.

67. About the body as a timekeeper more generally, see Dana Luciano's illuminating cultural history of grief, *Arranging Grief: Sacred Time and the Body in Nineteenth-Century America* (New York: New York University Press, 2007).

68. John Styles, *The Dress of the People: Everyday Fashion in Eighteenth-Century England* (New Haven: Yale University Press, 2007), 107. Styles writes: "For all except a few who worked in road transport, watches were not indispensable necessities when it came to the business of everyday timekeeping. What they did offer those affluent workers who could afford them was reasonably accurate, general-purpose timekeeping in a conveniently portable package."

69. Woloson, *In Hock*, 106.

70. James Parish Stelle, *The American Watchmaker and Jeweler, a Full and Comprehensive Exposition of all the Latest and Approved Secrets of the Trade . . . Also, Directions By Which Those Not Finding it Convenient to Patronize a Horologist May Keep Their Clocks in Order* (New York: Jesse Haney, 1868); H. F. Piaget, *The Watch*, 2nd ed. (New York: H. F. Piaget, 1868), viii. The American Antiquarian Society holds the first edition, which was published in 1860. Other period watch repair manuals include C. Hopkins, *The Watchmaker's and Jeweler's Hand-book* (Lexington, KY: Published for the author, 1866); F. Kelmo, *Kelmo's Watch-Repairers Hand-book; Being a Complete Guide to the Young Beginner in Taking Apart, Putting Together, and Thoroughly Cleaning the English Lever and Other Foreign Watches, and All American Watches* (Boston: A. Williams, 1869).

71. Joseph Gardner and Son, *Illustrated Catalogue of Watches and Description of Ball Clock in Their Window* (Boston, [1879]), Warshaw-WC, oversized, folder 6, box 203. The 1873 Elgin Almanac (box 1, Warshaw-WC) also advised that a watch should be cleaned and oiled every two years. According to Carlene Stephens, curator of technology at NMAH-SI, manufacturers made cheap watches in the 1880s that were meant to run without cleanings or repairs. However, well into the twentieth century, most watches required regular cleanings and repairs (personal correspondence in author's possession). About watch repair in the United States during the 1940s and 1950s, see Fred L. Strodtbeck and Marvin B. Sussman, "Of Time, the City, and the 'One-Year Guaranty': The Relations between Watch Owners and Repairers," *American Journal of Sociology* 61 (May 1956): 602–9.

72. Mark Twain, "My Watch: An Instructive Little Tale," in *Sketches New and Old* (Hartford, CT: Hartford Press, 1875), 11–15.

73. Styles, *Dress of the People*, 97, 99.

74. "Care of the Watch," *Kelmo's Watch-Repairer's Hand-Book*, 62–63.

75. David Glasgow, *Watch and Clock Making* (London: Cassell, 1885), 63–80; Henry G. Abbott, *Abbott's American Watchmaker and Jeweler: An Encyclopedia for the Horologist, Jeweler, Gold and Silversmith*, 5th ed. (1893; reprint Chicago: Geo. K. Hazlitt, 1895), 236–43.

76. Ed Ueberall and Kent Singer, "Elgin's Veritas Model," part 1, "The First Three Grades," *NB* 47 (April 2005): 210. Ueberall and Singer's quotation is from an advertisement in the *Jewelers' Circular*, August 7, 1901, 27. A 1906 ad for the same model watch claimed it would run for forty-two hours without being wound (211).

77. "The Exact Time," *NYT*, June 17, 1880.

78. Caroline Fairfield Corbin, *A Woman's Secret* (Chicago: Central Publishing House,

1867), 11; Charles H. Doe, *Buffets* (Boston: J. R. Osgood, 1875), 95; William Elder, *Periscopics; or, Current Subjects Extemporaneously Treated* (New York: J. C. Derby, 1854), 120.

79. Alfred W. Arrington, *The Rangers and Regulators of the Tanaha; or, Life among the Lawless* (New York: Robert M. DeWitt, 1856), 42–43. The theme of the gold watch as a sign and cause of profligacy is well-developed in T. S. Arthur's short story "It's Only a Dollar," in *Stories for Young Housekeepers* (Philadelphia: Lippincott, Grambo, 1854): 156–87.

80. The representative of a group of prosperous Brooklyn businessmen ironically said while presenting a gold watch to a retiring police officer: "Take it, wear it, and I hope it will always assist you to be on time to prevent a breach of the public peace" ("A Pleasant Affair," *BE*, December 23, 1880). Newspaper notices about the gifts of gold watches to long-serving and retiring employees occur infrequently between the 1880s and 1910s, but are numerous in the 1920s.

81. Carlene Stephens, *On Time: How America Has Learned to Live by the Clock*, (Boston: Bulfinch Press, 2002), 136–39. A Louisa May Alcott short story ends with the engagement of the heroine signified by a gift of a gold watch ("My Rococo Watch," in *Silver Pitchers: And Independence, A Centennial Love Story* [Boston: Little, Brown, 1876], 136–48).

82. Women owned nearly a third of all the gold watches Hoxie repaired. McCrossen, "Pocket Watches and Time Consciousness," 21, table 8.

83. See Stephens, *On Time*, 91, 92; O'Malley, *Keeping Watch*, 175–77.

84. An extant watch sales book of the Northampton jeweler B. E. Cook and Son covering the years 1884–93 suggests that about a hundred watches a year were sold. It is difficult to tell which watches were new and which ones were secondhand. The book is held in the Gere Collection, Historic Northampton.

85. McCrossen, "Pocket Watches and Time Consciousness," 15–21, esp. 18, tables 6 and 7.

86. Treasurer's Reports for 1881, 1883, 1884, 1887, 1888, and 1889, Waltham Records.

87. Landes, *Revolution in Time*, 317–20; Harrold, "Charles Rood and Henry Cain"; Harrold, "Economic Look at the American Watch Industry." See also tables 3 and 4 in Landes's *Revolution in Time* for the output of the Waltham Watch Company and the Elgin Watch Company. Waltham's output consistently exceeded a hundred thousand watches after 1876, hit one million in 1892, decreased more than 50 percent from that high point for the duration of the 1890s, and then hit one million again in 1902. Elgin's output too first exceeded a hundred thousand watches in 1875, and did not reach one million until 1900. Hampden Watch Company was started in Springfield, Massachusetts, as the Springfield Watch Company; by the mid-1880s it was the third leading watch manufacturer in the United States (Harrold, "Charles Rood and Henry Cain," 550–52).

88. Glasmeier, *Manufacturing Time*, 123–24; Landes, *Revolution in Time*, 317–20, tables 3–5. Michael Harrold suggests that dollar watches were available around 1890, and that by 1920 seven million "dollar watches" were sold every a year ("Economic Look at the American Watch Industry," 428).

CHAPTER FOUR

1. Patent 267,824, issued November 21, 1882, to John Wethered Bell, Conowingo, Maryland, for Watch-Hands; Patent 47,065 issued March 28, 1865, to A. W. Hall of New York, NY, assignor to B. W. Robinson and S. P. Chapin, for Improvement in Universal

Time-Pieces. About Fleming's patent applications, see Peter Galison, *Einstein's Clocks, Poincaré's Maps: Empires of Time* (New York: W. W. Norton, 2003), 117; Clark Blaise, *Time Lord: Sir Sandford Fleming and the Creation of Standard Time* (New York: Pantheon Books, 2000), 81–82. See also Mario Creet, "Sandford Fleming and Universal Time," *Scientia Canadensis* 14 (1990): 66–89.

2. Great Britain implemented standard time in 1848. See David Rooney and James Nye, "'Greenwich Observatory Time for the Public Benefit': Standard Time and Victorian Networks of Regulation," *British Journal for the History of Science* 42 (March 2009): 5–30.

3. "Bell's Improvement in Watch Hands," *Scientific American* 47 (December 16, 1882): 390. The article indicates that Bell patented the design in Canada, Great Britain, France, Belgium, Germany, Spain, Italy, and Austria, as well as the United States.

4. Michael O'Malley, *Keeping Watch: A History of American Time* (Washington, DC: Smithsonian Institution Press, 1990), 74, and "Tower Clocks," *Manufacturer and Builder* 1 (1869): 206.

5. The moment when prominent timekeepers were adjusted might have been akin to the dropping of a time ball. Michael J. Sauter, for instance, describes crowds of elite men gathering to watch the Saturday and Monday adjustment of the clock on Berlin's Academy of Sciences in the late eighteenth century ("Clockwatchers and Stargazers: Time Discipline in Early Modern Berlin," *AHR* 112 [June 2007]: 685–710).

6. Galison, *Einstein's Clocks*, 115.

7. Allen used the phrase "abolition of local time" in many letters; see for instance William F. Allen (WFA) to Major Sully, January 2, 1886, Letters Sent, 2:691–95, Allen Papers.

8. WFA to Mr. Riebenack, November 20, 1883, Letters Sent, 2:551–54, Allen Papers. Allen wrote his own history of the adoption of standard time, as did his son, great-grandson, and great-great-grandson: William F. Allen, "History of the Movement by Which the Adoption of Standard Time Was Consummated," *Proceedings of the American Metrological Society* 4 (1884): 25–50; William F. Allen, *Short History of Standard Time and Its Adoption in North America in 1883* (New York, 1904); John S. Allen, *Standard Time in America: Why and How It Came about and the Part Taken by the Railroads and William Frederick Allen* (New York, 1951); Frederick Warner Allen, "The Adoption of Standard Time in 1883, an Attempt to Bring Order into a Changing World" (undergraduate thesis, Yale University, 1969); Nathaniel Allen, "The Times They Are A-Changing: The Influence of Railroad Technology on the Adoption of Standard Time Zones in 1883," *History Teacher* 33 (2000): 241–56.

Other accounts that attribute the introduction of standard time to efforts made by Allen and the railroads include Robert E. Riegel, "Standard Time in the United States," *AHR* 33 (1927): 84–89; Sidney Withington, "Standardization of Time in Connecticut," *Railway and Locomotive Historical Society Bulletin* 46 (1938): 14–16; Sidney Withington, "Marking Time in 1883: A Contribution Made by the Railroads to Our National Welfare in Standardizing Time throughout the Country," *Connecticut Society of Civil Engineers Sixty-Seventh Annual Report* (1951), 120–33.

9. On the role of scientific organizations, see Carlene Stephens, "'The Most Reliable Time': William Bond, the New England Railroads, and Time Awareness in 19th-Century America," *Technology and Culture* 30 (1989): 1–24; Carlene Stephens, "The Impact of the Telegraph on Public Time in the United States, 1844–1893," *IEEE Technology and Society Magazine* 8 (March 1989): 4–10; Ian R. Bartky, "The Adoption of Standard Time,"

Technology and Culture 30 (1989): 25–56. On the history and meaning of the standardization of time in the United States more generally, see Barbara Liggett, "History of the Adoption of Standard Time in the United States, 1869–1883" (MA thesis, University of Texas, Austin, 1960); Eviatar Zerubavel, "The Standardization of Time: A Sociohistorical Perspective," *American Journal of Sociology* 88 (1982): 1–23; Carlene Stephens, *Inventing Standard Time* (Washington, DC: NMAH-SI, 1983); Orville Butler, "From Local to National: Time Standardization as a Reflection of American Culture," in *Beyond History of Science: Essays in Honor of Robert E. Schofield*, ed. Elizabeth Garber (Bethlehem, PA: Lehigh University Press, 1990): 249–65; Michael O'Malley, "Standard Time, Narrative Film and American Progressive Politics," *Time & Society* 12 (1992): 193–206; Mark M. Smith, *Mastered by the Clock: Time, Slavery, and Freedom in the American South* (Chapel Hill: University of North Carolina Press, 1997), 178–84; Ian R. Bartky, *Selling the True Time: Nineteenth-Century Timekeeping in America* (Stanford: Stanford University Press, 2000); Galison, *Einstein's Clocks*, 113–28.

About the standardization of time elsewhere, see Rooney and Nye, "'Greenwich Observatory Time'"; Eric Pawson, "Local Times and Standard Time in New Zealand," *Journal of Historical Geography* 18 (1992): 278–87; Blaise, *Time Lord*; Ian R. Bartky, *One Time Fits All: The Campaigns for Global Uniformity* (Stanford: Stanford University Press, 2007); Adam Barrows, *The Cosmic Time of Empire: Modern Britain and World Literature* (Berkeley: University of California Press, 2011), 1–52.

10. The motif of the "technological fix" informs much of this chapter. The term comes from Lisa Rosner, ed., *Technological Fixes: How People Use Technology to Create and Solve Problems* (New York: Routledge, 2004).

11. Ian R. Bartky, "The Invention of Railroad Time," *Railroad History* 148 (1983): 12–22, esp. 18–20.

12. "Isham's City Time" on No. 17 clock (Titusville, PA): box 8, 1:26 (1871), HC; "Isham's City Time" on No. 36 regulator (Titusville, PA): box 8, 1:37 (1872), HC; "S. A. Orth City Time" (Cleveland), box 8, 1:49 (1872), HC; "Railroad Time H. Cooper" on No. 36 Regulator (Wellsville, OH): box 8, 1:63 (1872), HC; "M. Parker & Co. T. W. & W. R. R. Time" on No. 36 Regulator (Danville, IL): box 8, 1:126 (1873), HC; "A. R. Everett City Timekeeper Atlanta, Geo." on No. 60 regulator: box 8, 3:31 (1877). HC; "E. L. Barlow & Son City Time" on No. 61 regulator (Georgetown, KY): box 8, 3:60 (1877), HC; "City & Train Time" on No. 60 Regulator (Charlotte, MI): box 8, 3:68 (1877), HC; "Railroad Time" on No. 3 Regulator (Norwalk, OH): box 8, 6:103 (1882), HC.

13. Trade card, "C. W. Platt, Repairer and Adjuster of Fine Watches, Chronometers, Repeaters and Chronographs," [ca. 1880s], entry 8, box 3, USNO-Records. The card came with two letters from Platt to the USNO asking about how to use his transit instrument. He writes: "I am very anxious to be correct. I have a good chronometer & regulator and know what it is to have and keep correct time" (letter of September 29, 1886, entry 8, box 4, USNO-Records).

14. Forty-one clocks meant to serve as time standards were ordered from E. Howard and Seth Thomas between 1871 and November 18, 1883.

15. "Platts Standard Time" (Delaware, OH), box 8, 1:62 (1872), HC; "True Time E. Gray Co." (Akron, OH), box 8, 1:49 (1872), HC; "J. M. Heard Meridian Time" (Cleveland), box 8, 1:49 (1872), HC; "W. Wilson McDrew Standard Time" (Cincinnati): box 8, 1:96 (1873), HC; "Correct-time" (Helena, Montana), box 8, 2:81 (1875), HC; "Standard Time Cowell Bros." (Cleveland): book B, 199 (1878), ST.

16. "The True Time," undated news clipping in Chancellor's Scrapbooks, Scrapbook

1867–77, Washington University Archives, emphasis in original. Thanks to Adam Arenson for sharing this source with me. Washington University ordered a E. Howard astronomical clock for its new observatory in 1881 (box 8, 5:118 [1881], HC).

17. The Harvard Observatory began to develop its time service in 1871 after it fell into disuse during the 1860s. Under astronomer Leonard Waldo, who was hired in 1877 to run the service, the focus was on precision, which suited New England's owners of fine watches. See Galison, *Einstein's Clocks*, 107–12. Charles Teske's efforts in Hartford in 1878 are also discussed by Galison (110–11).

18. Edward Vail's order and the French Atlantic order can be found respectively in box 8, 1:25 (1871) and 4:86 (1880), HC. Similar orders for clocks with extra minutes hands include box 8, 2:114 (1875) and 5:164 (1882). About clocks with dual minute hands, see Clive N. Ponsford, *Time in Exeter: A History of 700 Years of Clocks and Clockmaking in an English Provincial City, with Details of More than 300 Former Makers* (England: Headwell Vale Books, 1978), 11–12.

19. Box 8, 5:90, 91, 120, 123, 132, 137, 167, and box 10, 3:201, HC. The first order was placed August 16, 1881, the last order February 20, 1882.

20. Quotation from Bartky, *Selling the True Time*, 82; see also Galison, *Einstein's Clocks*, 94–97.

21. Mayo's order was dated October 27, 1871 (box 8, 1:19 [1871], HC). At the end of 1877, the Dearborn Observatory supplied the time to six Chicago jewelers, four Chicago railways, and the Chicago Board of Trade (Galison, *Einstein's Clocks*, 112).

22. Peter Putnis, "News, Time and Imagined Community in Colonial Australia," *Media History* 16 (2010): 153–70; Terhi Rantanen, "The Globalization of Electronic News in the 19th Century," *Media, Culture and Society* 19 (1997): 605–20; Tim Edensor, "Reconsidering National Temporalities: Institutional Times, Everyday Routines, Serial Spaces and Synchronicities," *European Journal of Social Theory* 9 (2006): 525–45. About the role of simultaneity in fostering nationalism, see Benedict Anderson, *Imagined Communities: Reflections on the Origin and Spread of Nationalism*, new and rev. ed. (New York: Verso, 2006), esp. 24–26.

23. *Evening Post* clock, box 8, 2:93 (1875), HC; *New York Times* clock, book A, 85 (1877), ST; *New York Clipper* clock, book B, 270 (1879), ST; *Tribune* clock, E. Howard Clock Company, *Catalogue of Regulators, Office and Bank Clocks* (1901). About newspaper row and the *Tribune* building, see Robert A. M. Stern, Thomas Mellins, and David Fishman, *New York 1880: Architecture and Urbanism in the Gilded Age* (New York: Monacelli Press, 1999), 402–6.

24. Photograph, Lincoln Market, Prints Department, LCP; Pennsylvania Railroad clocks, box 8, 1:1, 68 (1871–72), HC; Guarantee Trust building clock, box 8, 2:33 (1874), HC; Wanamaker's 1880 order for an hour striking clock, book B, 280 (1880), ST.

25. Arthur H. Frazier, "Henry Seybert and the Centennial Clock and Bell at Independence Hall," *Pennsylvania Magazine of History and Biography* 102 (1978): 47; Alain Corbin, *Village Bells: Sound and Meaning in the 19th-Century French Countryside*, trans. Martin Thom (1994; reprint, New York: Columbia University Press, 1998), 267, also 141–48, 264–83. Seth Thomas received some orders for bells with inscriptions with the names of commissioners or clock committee members, see book E, 84–85, 102–3 (1892); book H, 35–36 (1903), 160 (1905); book I, 139 (1909). A Troy, Indiana, clock committee requested an inscription for its 2,000-pound bell with their names and the command "Tell the Time to the Trojans" (book D, 35 [1888], ST). Seth Thomas also fulfilled orders

for bells inscribed with memorials, such as the following meant for a 1,013-pound bell to hang from the Polk County Court House in Bolivar, Missouri: "Presented to the People of Polk County by Virginia Bushnell Townsend in Memory of her husband Orra Milton Townsend 1907" (book H, 267 [1907], ST). For the Lutheran Orphans Home in Topton, Pennsylvania, a daughter ordered a bell inscribed with a memorial to her parents (book I, 183 [1910], ST).

26. Frazier, "Centennial Clock," 54; Untitled article, *PT*, December 31, 1893; Wilfred Isidor (curator of the Bureau of City Property) to bureau chief, September 22, 1913, folder 19, box 19, ser. 1, subser. A, BCP-IH; curator of Independence Hall to Mrs. Joseph Foisy, January 26, 1924, folder 33, box 19, ser. 1, subser. A, BCP-IH. The city gave the state house's old clock, dilapidated dials, and 4,000-pound bell to nearby Germantown (F. M. Ritter [director of Independence Hall] to A. S. Eisenhower [chief of the Bureau of City Property], May 20, 1898, typescript in *Documents Relating to the Physical History of Independence Hall*, vol. 2, IHA).

27. Since 1876, replicas of the Liberty Bell have accumulated, ranging from thimble-sized, half-pound souvenirs to the 13,000-pound bell hung that year on Independence Hall to a 20,000-pound bell called "the World Freedom Bell" hung in Berlin in 1950 (Gary Nash, *The Liberty Bell* [New Haven: Yale University Press, 2010], 64–65, 91–92). In 1950, the Treasury Department gave each state a replica of the Liberty Bell. Today twenty-nine state capitols still have it on display. See Charles T. Goodsell, *The American Statehouse: Interpreting Democracy's Temple* (Lawrence: University Press of Kansas, 2001), 101–9.

28. Bartky, *Selling the True Time*, 50–57. After the 1850s, telegraph wires began to connect bells with time services provided by astronomical observatories and later by companies that specialized in the sale of time signals.

29. Francis A. Walker, ed., *International Exhibition, 1876: Reports and Awards, Group XXV* (Philadelphia: J. B. Lippincott, 1878), 53–28, 154–64, 200–201; James C. Watson, *American Watches: An Extract from the Report of Horology at the International Exhibition at Philadelphia, 1876* (New York: Robbins and Appleton, 1877), held by the Hagley Museum and Library; *Souvenir of the Centennial Exhibition: Connecticut's Representation at Philadelphia, 1876* (Hartford: Geo. D. Curtis, 1877), 98; James D. McCabe, *An Illustrated History of the Centennial Exhibition* (Philadelphia: National Publishing, 1876), 354–55, 434, 447. For a sense of the extent of the mechanical timepieces on display at the Centennial Exhibition, see the Centennial Exhibition Digital Collection, FLP, http://libwww.freelibrary.org/cencol/ (accessed June 21, 2011).

30. USNO astronomical clock order, box 8, 1:122 (1873), HC; Government Hospital for the Insane orders, box 8, 1:4 (1871), and 2:15 (1874), HC; Treasury Department orders, box 8, 1:89 (1873)and 3:64 (1877), HC. An assistant quartermaster for the army placed an order with Seth Thomas for a No. 16 striking tower clock for two six-foot dials with gilt hands and numerals for San Antonio's US Depot Tower (book B, 379 [1882], ST). A few years later, an order for a third dial was placed (book C, 103 [1886], ST). Seth Thomas and E. Howard supplied San Antonio with other clocks as well: a one-dial clock for a shop (book B, 294 [1880]); a post clock for a jeweler (box 8, 6:16 [1882], HC); eight No. 70 clocks for the US Courthouse and Post Office (box 9, 2:235 [1890], HC); and new clockworks for a tower clock (book H, 286 [1907], ST).

31. Bartky, *Selling the True Time*, 47–74, 106–14; O'Malley, *Keeping Watch*, 74–75, 88. For contemporary accounts of the Western Union time ball, see "Standard Time: The New Western Union Time Ball in Operation," *Journal of the Telegraph* 10 (October

1877): 289–90; Edward S. Holden, "On the Distribution of Standard Time in the United States," *Popular Science Monthly* 11 (1877): 175–82; *Scientific American* 39 (November 1877): 335, 337.

32. Winslow Upton, "Information Relative to the Construction and Maintenance of Time Balls," *Professional Papers of the Signal Service* 5 (Washington, DC: GPO, 1881). The building was on the corner of Milk and Devonshire Streets. William Bond recorded watching the ball drop for the first time in his daybook (entry of May 8, 1878, Daybooks 1857–1903, box 15, Bond Records).

33. H. S. Pritchett, "The Kansas City Electric Time Ball," *Kansas City Review of Science* 5 (April 1881): 720–24, quotation 723. For a list of time ball locations, see Bartky, *Selling the True Time*, 159, appendix.

34. Cleveland Abbe, "Report on Standard Time," *Proceedings of the American Metrological Society* 2 (1880): 17–45. About Abbe's efforts to implement standard time, see Bartky, "Adoption of Standard Time," 34–39.

35. W. B. Hazen, chief signal officer of the United States, to F. A. P. Barnard, president of the American Metrological Society, March 1, 1881, reprinted in William F. Allen, *History of the Adoption of Standard Time* (np, nd), box 2, folder 4, Allen Papers. The memo listing signal service stations amenable to time balls was enclosed with the letter.

36. Bartky, "Adoption of Standard Time," 43–48.

37. WFA to W. B. Hazen, September 27, 1883, Letters Sent, 1:240, Allen Papers. For an account of the adoption of standard time in Boston and a subsequent court case adjudicated by Justice Oliver Wendell Holmes, see Arnold A. Lasker, "A Time to Arrest and a Time to Release," *New England Quarterly* 58 (March 1985): 83–87. It may be that there was some resistance to adopting the 75th-meridian time because the Boston area's flourishing "culture of precision" demanded "accuracy to tenths if not hundredths or even thousandths of a second" in accord with the port town's exact meridian (Galison, *Einstein's Clocks*, 112). Galison argues that the Harvard Observatory "cultivated precision mania." It is worth considering how the Waltham Watch Company and the E. Howard Company also furthered this mania for precision.

38. WFA to W. H. Barnes, September 10, 1883, and WFA to Charles Francis Adams Jr., September 13, 1883, Letters Sent, 1:170–72 and 1:190–91, Allen Papers.

39. In Pickering's absence, Allen proceeded to write to astronomers asking them to help persuade Pickering and the president of Harvard College. From Letters Sent, Allen Papers, see WFA to W. B. Hazen, September 27, 1883, 1:240; WFA to F. A. P. Barnard, September 27, 1883, 1:242; and WFA to Cleveland Abbe, September 29, 1883, 1:259, and October 1, 1883, 1:272.

40. WFA to W. E. Barnes, October 9, 1883, Letters Sent, 1:328–29, Allen Papers; J. Rayner Edmands to WFA, November 11, 1883 (post card), and November 11, 1883, Letters Received, 4:47 and 4:51, Allen Papers, emphasis in original.

41. WFA to James J. Eckert, September 14, 1883, and WFA to Charles E. Pugh, September 28, 1883, Letters Sent, 1:195–96 and 1:249, Allen Papers.

42. From Letters Sent, Allen Papers, see WFA to E. S. Brown (general superintendent of New York, Lake Erie and Western Rail Road) and WFA to J. M. Toucey (general superintendent of New York Central and Hudson River Rail Road), September 29, 1883, 1:260, 264; WFA to James Hamblet, October 1, 1883, 1:269.

43. WFA to Franklin Edson, October 19, 1883, Letters Sent, 2:372–74, Allen Papers.

44. WFA to J. A. Filmore, September 27, 1883, Letters Sent, 1:244–45, Allen Papers. About the powerful pull Greenwich exerted, see Barrows, *Cosmic Time of Empire*, esp.

75–128. Deirdre Murphy has argued that the introduction of a national system of railroad standard times in 1883 underscored American political tendencies toward xenophobia, which in that decade were expressed through disfranchisement of African American men and the passage of the Chinese Exclusion Act in 1882. Her argument, while not fully developed, is thought-provoking. See Deirdre Murphy, "The Look of a Citizen: Representations of Immigration in Gilded Age Painting and Popular Press Illustration" (PhD diss., University of Minnesota, 2007), chap. 3.

45. WFA to Charles Pugh, November 13, 1883, Letters Sent, 2:494, Allen Papers.

46. WFA to Prof. E. Kirby, November 28, 1883, Letters Sent, 2:584–85, Allen Papers.

47. WFA to Admiral J. Rowan, October 6, 1883, Letters Sent, 1:304–7, and R. W. Shufeldt to WFA, October 9, 1883, Letters Received, 2:101, Allen Papers. Allen expressed unawareness of the USNO's operations and staffing, thus the confusion in his addressing the wrong man as USNO superintendent.

48. W. F. Sampson to WFA, October 20, 1883, Letters Received, 2:157, Allen Papers.

49. Bartky, "Adoption of Standard Time," 39–40. The 1881 plan would have made observatory time the official time of Washington, DC, and of the federal government. It never came up for a vote. In 1882, the US Postal Service began to follow a common time (Galison, *Einstein's Clocks*, 123).

50. WFA to assistant general manager of the Pennsylvania Railroad Company, August 21, 1883, Letters Sent, 1:138, Allen Papers, emphasis added. Clark Blaise, the biographer of Sandford Fleming, quotes an early time reformer, Charles Dowd, as writing: "The traveler's watch was to him but a delusion; clocks at stations staring each other in the face defiant of harmony either with one another or with surrounding local time and all wildly at variance with the traveler's watch, baffled all intelligent interpretation" (Blaise, *Time Lord*, 97).

51. William F. Allen, editorial, *Travelers' Official Railway Guide*, September 1883), no page numbers; WFA to Prof. E. Kirby, November 28, 1883, and WFA to John Adams, October 4, 1883, Letters Sent, 2:584–85 and 1:290, Allen Papers.

52. Bartky, "Adoption of Standard Time," quotes Allen about hearing the bells chime (49). Allen's letters describe his happiness about seeing the time ball drop. See the following correspondence from Letters Sent, Allen Papers: WFA to D. H. Bates, November 17 and November 20, 1883, 2:527 and 2:532; WFA to George A. Dodman, November 20, 1883, 2:533; and WFA to E. K. Moore, November 27, 1883, 2:580–81.

53. The 60th meridian set the time for a fifth time zone in easternmost Canada.

54. "Set Back Nine Minutes," *CDN*, November 19, 1883. Other newspapers from across the country made similar reports. In St. Joseph, Missouri, for example, after "the great clock in the tower over the Saxton National Bank struck twelve o'clock," its "hands were turned backward twenty minutes, and the bell now rings on standard time." The report continued, "Saxton's clock was utilized by scores of watch owners this morning who were setting their time-keepers to the new time" (*St. Joseph Evening News*, November 19, 1883).

55. "Time by Telegraph," *Railroad Gazette* 14 (November 17, 1882): 713.

56. Cleveland's Brunner Brothers placed their clock order November 14, 1883 (box 10, 3:53, HC). That the Ohio city would stick with its local time seemed such a strong likelihood that the jewelers bought the $240 clock on credit.

57. In September 1884, the Pendleton County Courthouse, in Falmouth, Kentucky,

placed its order for a No. 1 striking clock to run four four-foot dials and strike a four-hundred-pound bell (box 10, 3:95 [1884], HC). Other E. Howard orders for clocks with extra minute hands were placed around the same time from Pittsburg and from Cleveland (box 10, 3:382, 390 [1884], HC).

58. James C. Scott, *Seeing Like a State: How Certain Schemes to Improve the Human Condition Have Failed* (New Haven: Yale University Press, 1998), 11–71, quotations 25, 55.

59. John Rodgers to WFA, February 20, 1882, entry 4, 4:552–55, USNO-Records. The letter is also in Letters Received, 1:10, Allen Papers.

60. John W. Bell, "Standard Time," undated handwritten manuscript, 4–6, 10, box 2, folder 11, Allen Papers.

61. Frederick T. Newberry, "Standard Time," handwritten manuscript (ca. March–August 1882), in Letters Received, 1:17, Allen Papers.

62. Charles F. Dowd, *The Railway Superintendent's Standard Time Guide* (Saratoga Springs, NY, 1877), 4; Charles F. Dowd, *The Traveler's Railway Time Adjuster, with over 800 Time Corrections* (Saratoga Springs, NY, 1878), 3; Charles F. Dowd, *System of Time Standards Illustrated with Map* (Saratoga Springs, NY, 1884), no page numbers. On Dowd's time-difference charts, see O'Malley, *Keeping Watch*, 79–82.

63. WFA to Bailey, Banks and Biddle (Philadelphia) and WFA to J. B. Mayo (Chicago), December 16, 1882, Letters Sent, 1:54, Allen Papers.

64. WFA to John Bell, March 31, 1883, April 4, 1883, and April 20, 1883, Letters Sent, 1:66, 1:78, and 1:88–89. Bell describes the watch in a letter to Allen dated April 3, 1883, Letters Received, 1:77–79, Allen Papers.

65. Bell, "Standard Time," 1–3. Before the standardization of time, regional railways in the United States followed at least fifty different time standards. See Carlene Stephens, *On Time: How America Has Learned to Live by the Clock* (Boston: Bulfinch Press, 2002), 111.

66. Jack Linahan, "Watch and Clock Collector, Henry Ford," *NB* 46 (October 2004): 579–85, esp. 581.

67. Bell describes his encounter with the German watchmaker in a letter to Allen dated December 18, 1882, Letters Received, 1:60, Allen Papers, emphasis in original.

68. See the following correspondence from Bell to WFA in Letters Received, Allen Papers: January 6, 1883 (1:62); February 12, 1883 (1:65); February 21, 1883 (1:68); June 4, 1883(1:100).

69. Bell to WFA, January 24, 1883, Letters Received, 1:64, Allen Papers; advertisement for Bell's innovation in *The Official Railway Guide: North American Freight Service* (Philadelphia, 1883), 482.

70. Bell to WFA, February 12, 1883, Letters Received, 1:65, Allen Papers.

71. Bell, "Standard Time," 2. About Admiral John Rodgers's views concerning sun time, see the following letters he wrote in 1881 and 1882: Rodgers to W. B. Hazen (chief signal officer), June 11, 1881, and Rodgers to WFA, February 20, 1882, entry 4, 4:471–72, 552–55, USNO-Records.

72. Wolfgang Schivelbusch, *Disenchanted Night: The Industrialization of Light in the Nineteenth Century*, trans. Angela Davies (1983; reprint, Berkeley: University of California Press, 1995); Peter Baldwin, *In the Watches of the Night: Life in the Nocturnal City, 1820–1930* (Chicago: University of Chicago Press, 2012); Daniel Freund, *American Sunshine: Diseases of Darkness and the Quest for Natural Light* (Chicago: University of Chicago Press, 2012).

73. Bell quotes Admiral John Rodgers in "Standard Time," 1. In a letter to Allen, Rodgers opens: "The sun regulates all life upon the earth, vegetable and animal" (Rodgers to WFA, February 20, 1882, entry 4, 4:552–55, USNO-Records.)

74. Galison, *Einstein's Clocks*, 125.

CHAPTER FIVE

1. "Four Faced, But Honest," *BE*, March 2, 1890, 8. A similar article about Chicago punned about the city's clocks being two-faced, four-faced, and even six-faced: "City of Big Tower Clocks," *Washington Post* (byline from the *Chicago Tribune*), March 8, 1896.

2. "No Tick There," *Daily Ardmoreite* (Ardmore, OK), December 13, 1901; "What Time Is It?" *NA*, May 4, 1899, clipping, Perkins Scrapbook. "No Tick There" shares an anecdote about a salesman who, as a result of ignoring his misgivings about a town clock, missed a train. It describes a type it calls "diffident men," who are "the sort who don't compare watches with the town clock and tell everyone for a block around that the clock is seven minutes off." If these men do compare the two, they do so "in a furtive way."

3. Waltham Clock Company advertisement for electric time system in 1899 catalog, quoted in Andrew H. Dervan, "Waltham Clock Company," *NB* 47 (April 2005): 205.

4. Out of sixty orders for clocks placed with Seth Thomas and E. Howard by railroads between 1871 and 1883, only six were for exterior clocks. Out of thirty-eight orders railroads placed for clockworks with Seth Thomas and E. Howard between 1884 and 1889, seventeen were meant to operate dials on the exterior of depots and terminals. Out of sixty-six orders placed during the 1890s: twenty-five were for clockworks to operate a single large dial hung inside or outside the depot, twenty-two were for three- or four-dial tower clocks, two were for post clocks, and the remaining seventeen were for regulators with twelve or twenty-four-inch dials. The orders placed between 1902 and 1911 were all for clocks hung on the exterior of railroad stations.

In 1871, when New York City's "Grand Central Depot" opened at Forty-Second Street and Fourth Avenue, it was not yet in the center of Manhattan. When Chicago's "Grand Central Passenger Station" opened in 1891, the city's other railroad terminals were all located near the river, far from commercial and residential districts. St. Louis's Union Station opened in 1894 in a part of town that at the time was considered remote. The District of Columbia's first railroad station, not much more than a shanty, was at Pennsylvania Avenue and Second Street NW. In 1850, a new depot with a clock tower went up far from the Capitol. In 1901, Union Station opened; it had no exterior clocks. A period guide to building railroad structures described more than two dozen clock towers that were built in the second half of the 1880s alone (Walter Berg, *Buildings and Structures of American Railroads* [New York: John Wiley and Sons, 1893], 303–4, 311, 319, 322, 336, 363, 371–72, 374–76, 385–87, 401, 409, 424). See Carl W. Condit, *The Port of New York*, vol. 1, *A History of the Rail and Terminal System from the Beginnings to Pennsylvania Station* (Chicago: University of Chicago Press, 1980), 86; *Grand Central Passenger Station, Chicago* (Chicago: Exhibit Publishing Company, 1891), no page numbers, DL-SMU; Osmund Overby, "A Place Called Union Station: An Architectural History of the St. Louis Union Station," in *St. Louis Union Station: A Place for People, a Place for Trains* (St. Louis: St. Louis Mercantile Library, 1994), 60, 70; Washington Topham, *First Railroad into Washington and Its Three Depots* (Washington, DC: Columbia Historical Society, 1925), DL-SMU.

5. The Broad Street Station opened in 1881; its Gothic clock tower is visible in an 1882 photograph held by the Prints Department, LCP. In addition to the four dials on the clock tower, an illuminated terra cotta clock dial was installed on its exterior some years later (box 9, 1:279–80 [1894], HC).

6. Post clocks ordered from E. Howard (box 9, 2:192, 225, 230 [1890], HC).

7. Carol Greenhouse, *A Moment's Notice: Time Politics across Cultures* (Ithaca: Cornell University Press, 1996).

8. Wu Hung, "Monumentality of Time: Giant Clocks, the Drum Tower, the Clock Tower," in *Monuments and Memory, Made and Unmade*, ed. Robert S. Nelson and Margaret Olin (Chicago: University of Chicago Press, 2003), 108, 110, 113; see also 113–27.

9. Alain Corbin, *Village Bells: Sound and Meaning in the 19th-Century French Countryside*, trans. Martin Thom (1994; reprint, New York: Columbia University Press, 1998), 8–14, 20–23; Gerhard Dohrn-van Rossum, *History of the Hour: Clocks and Modern Temporal Orders*, trans. Thomas Dunlap (1992; reprint, Chicago: University of Chicago Press, 1996), 139–40.

10. Chris McKay, *Big Ben: The Great Clock and the Bells at the Palace of Westminster* (Oxford: Oxford University Press, 2010), 2–7.

11. Graeme Davison argues that time awareness first changed in Australia in the 1820s and 1830s when "colonial officials sough to reinforce the dissolving structures of penal society by the introduction of new clock-based systems of authority, and erected a rudimentary system of public time" (*The Unforgiving Minute: How Australia Learned to Tell the Time* [Melbourne: Oxford University Press, 1993], 1–2; see also 18, 20–21).

12. Yujin Yaguchi, "American Objects, Japanese Memory: 'American' Landscape and Local Identity in Sapporo, Japan," *Winterthur Portfolio* 37 (Summer/Autumn 2002): 93–121.

13. Sultan Abdülhamid II quoted in Mehmet Bengü Uluengin, "Secularizing Anatolia Tick by Tick: Clock Towers in the Ottoman Empire and Turkish Republic," *International Journal of Middle Eastern Studies* 42 (2010): 20; see also 17–36. See as well Avner Wishnitzer, "A Comment on Mehmet Bengü Uluengin, 'Secularizing Anatolia Tick by Tick: Clock Towers in the Ottoman Empire and Turkish Republic,'" *International Journal of Middle Eastern Studies* 42 (2010): 537–45.

14. Qin Shao, "Space, Time, and Politics in Early Twentieth Century Nantong," *Modern China* 23 (January 1997): 99–129; Wen Hsin Yeh, "Corporate Space, Communal Time: Everyday Life in Shanghai's Bank of China," *AHR* 100 (February 1995): 97–122, esp. 99–102; Jeffrey N. Wasserstrom, "A Big Ben with Chinese Characteristics: The Customs House Tower as Urban Icon in Old and New Shanghai," *Urban History* 33 (May 2006): 65–84. With the founding of the Republic of Turkey in 1923, Atatürk oversaw the construction of several prominent clock towers in the Art Deco style, while at the same time some of Turkey's older clock towers, reminiscent of Ottoman rule and sometimes mistaken for church towers, were demolished or left in states of disrepair (Uluengin, "Secularizing Anatolia Tick by Tick," 26–30).

15. During the first half of the nineteenth century, "most manifestations of government presence were small or makeshift" (Lois A. Craig, *The Federal Presence: Architecture, Politics, and Symbols in U.S. Government Building* [Cambridge: MIT Press, 1984], 51).

16. Mary P. Ryan, "'A Laudable Pride in the Whole of Us': City Halls and Civic Materialism," *AHR* 105 (2000): 1131–70, quotation 1168.

17. Henry-Russell Hitchcock and William Seale, *Temples of Democracy: The State Capi-*

tols of the USA (New York: Harcourt Brace Jovanovich, 1976), 7, 63, 125–27, 266. An exception to the absence of clocks on state capitols is the second Alabama State Capitol (1851), which incorporated a four-faced clock sited on a cupola above its portico (125–27). Hitchcock and Seale note that the erection of the dome on the United States Capitol, completed in 1866, reinforced the prevalence of this design among state capitols. Deliberately conveying a commitment to timelessness, the US Capitol Building has no exterior clocks. About state capitols, see also Charles Goodsell, *The American Statehouse: Interpreting Democracy's Temple* (Lawrence: University of Kansas Press, 2001).

18. Thomas Allen, *A Republic in Time: Temporality and Social Imagination in Nineteenth-Century America* (Chapel Hill: University of North Carolina Press, 2008), 13–14. Another scholar of American literature, Lloyd Pratt, calls attention to the variety of temporalities evoked in different genres of nineteenth-century American literature, which he argues manifested a deep degree of temporal heterogeneity that inhibited the formation of national and ethnic identities. Allen, on the other hand, argues that temporal heterogeneity and layering fostered the development of national identity. See Pratt, *Archives of American Time: Literature and Modernity in the Nineteenth Century* (Philadelphia: University of Pennsylvania Press, 2009), 3 and 5.

19. Bob Frishman, "Mathew Brady's Clock," *NB* 44 (October 2002): 605–8. Out of seven thousand portraits, sixty featuring the clock were identified (605). They included portraits of George Custer, Andrew Stephens (the vice president of the Confederate States), Salmon Chase (chief justice of the Supreme Court), Schuyler Colfax (Ulysses Grant's vice president), and others.

20. W. F. Allen, *Railroad Gazette* 16 (April 14, 1884): 254–55.

21. Forty-eight clocks explicitly meant to serve as time standards were ordered from E. Howard and Seth Thomas between November 18, 1883, and the end of 1909; another forty-one were ordered between 1871 and November 14, 1883. Jewelers placed 43 percent of all orders for clocks meant to serve as time standards. E. Howard orders for clocks engraved "standard time" placed after 1883 include box 9, 1:10, 17, 123, 222, 223 (1888) and 2:55, 200 (1890); box 10, 1:184, 196 (1896), 2:11, 32 (1897), and 3:73, 82, 90, 99 (1884). Seth Thomas orders include book C, 113 (1886); book E, 90 (1892); book H, 122–24, 131, 211, 213, 276, 293 (1905–7); book I, 15, 44, 57, 59, 71, 79, 124, 142 (1908–9).

22. Scooler placed the orders October 30, 1883 (box 10, 3:50, 51 [1883], HC). The previous year he ordered from E. Howard a tower clock for two six-and-a-half foot dials that would strike a 750-pound bell (box 8, 6:67 [1882], HC). Perhaps this order was placed for the New Orleans Cotton Exchange Building that was finished in 1883. In 1892, Scooler ordered a Seth Thomas striking tower clock for the Plaquemines Parish Courthouse (book E, 123 [1892], ST). The clock on the Sugar Exchange is visible in a photograph taken between 1882 and 1900 by George F. Mugnier held by the Louisiana State Museum http://louisdl.louislibraries.org/u?/GFM,97 (accessed March 12, 2012).

23. The Linde order was placed January 17, 1884 (box 10, 3:61 [1884], HC); the Nordman Brothers' order was placed April 16, 1884 (box 10, 3:73 [1884], HC); and Davidson, Leyson and McCune's order was placed April 21, 1890 (box 9, 1:117 [1890], HC).

24. Self-identified jewelers placed 107 orders from E. Howard and Seth Thomas for clocks intended for use in association with their shops; undoubtedly there were many more jewelers among E. Howard's and Seth Thomas's customers, but the account books did not identify them by their trade.

25. Joseph Jessop, "The Jessop Street Clock," *Journal of San Diego History* 33 (Winter 1987), http://www.sandiegohistory.org/journal/87winter/clock.htm (accessed March 6, 2012).

26. E. J. Crane advertisement, *Richmond Planet*, February 23, 1895, 4. This advertisement, complete with an engraved bust of Crane, ran throughout 1895 and 1896. Crane moved to Richmond from Savanna, Georgia, in 1892. In Savannah he worked as a jeweler and watchmaker, as well as the editor and publisher of the paper *Labor Union*. In 1898 he was nominated to run for the Richmond Common Council on the Republican ticket ("Jackson Ward Ticket," *Richmond Planet*, May 14, 1898).

27. Detroit Publishing Company photograph of Mulberry Street, New York, ca. 1900–1910, LC-D418-9350, LC-PPD. Another view of a shop selling watches and clocks to an immigrant neighborhood is seen in a photograph of an Italian shop on New York City's Mott Street, taken in 1912, LC-B2-3982, LC-PPD.

28. Ryan, "City Halls and Civic Materialism," 1168.

29. The dates for the construction of city halls are found in William Lebovich, *America's City Halls* (Washington, DC: Preservation Press, 1984), 64–65, 68–69, 82, 88–89, 96, 102, 113, 128.

30. Owing to a number of factors, including increased state involvement in urban affairs, buildings housing both county courthouses and city halls were typical of the second half of the nineteenth century (Lebovich, *America's City Halls*, 20).

31. "The New Town Clock," *Hampshire Gazette*, June 4, 1878. E. Howard supplied the clock. In 1906, the sons of prominent jeweler and silversmith Benjamin E. Cook provided the funding to illuminate the First Church's clock tower as a memorial to their father ("Illuminated Clock Will Be Secured," *Hampshire Gazette* January 25, 1906). See city orders for striking tower clocks that would be erected on local churches placed by the Hackensack, New Jersey, Committee on Town Clock (box 9, 2:76 [1890], HC); the mayor of Portsmouth, New Hampshire (box 9, 3:122 [1893], HC); and the West Rindge, New Hampshire, Board of Selectmen (box 10, 1:85 [1895], HC). See also newspaper accounts, such as one noting that the town clock for Huntington, Long Island, was installed "in the new Methodist clock tower" (*BE*, October 16, 1900). Additional references to town clocks on churches can be found in town records. Mark M. Smith addresses the incidence of public clocks found on churches in the colonial and antebellum South. He also addresses the "coexistence of sacred, natural, and clock time," which city clocks on churches clearly highlights (*Mastered by the Clock: Time, Slavery, and Freedom in the American South* [Chapel Hill: University of North Carolina Press, 1997], 17–20, 42–48).

32. J. W. Benson, *Time and Time-Telling* (1875), quoted in Frederick M. Shelley, *Early American Tower Clocks: Surviving American Tower Clocks from 1726 to 1870* (Columbia, PA: NAWCC, 1999), ix.

33. The order was placed June 5, 1893 (Box 9, 3:109 [1893], HC). Zion's Evangelical Reformed Church and St. Ames Church of Buffalo, New York, also ordered tower clocks for installation in time for July Fourth (box 10, 1:60, 89 [1895], HC).

34. Counties in Georgia and Texas accounted for more than half of the orders, but orders from counties in the remaining Southern states were evenly distributed.

35. By 1875, Georgia had 137 counties; 9 were added in August 1905 (none of which ordered E. Howard tower clocks that year); and today there are 159 counties. Photographs of extant Georgia county courthouses show that many have prominent clock towers.

36. The earliest Seth Thomas order is from 1878, when Brazos County ordered a striking tower clock (book B, 194 [1878], ST); the last order was placed in 1892 for the

Haskell County Courthouse (book E, 86–87 [1892], ST). The earliest E. Howard order is from 1884, when Bastrop County ordered a striking tower clock (box 10, 3:67 [1884], HC); the last order is from 1897, when Brazoria County ordered a mechanism to run four four-foot dials and ring a eight-hundred-pound bell (box 10, 2:30 [1897], HC). The Seth Thomas accounts also show that four more Texas counties ordered striking tower clocks in the first decade of the twentieth century; Edna County, Foard County, Kinney County and Jefferson county (book H, 178 [1905]; book I, 184 [1910]; book I, 196 [1910]; book I, 208 [1910]). See also Mavis P. Kelsey and Donald H. Dyal, *The Courthouses of Texas* (College-Station: Texas A&M University Press, 1993), 22–24.

37. Antoinette Lee, *Architects to the Nation: The Rise and Decline of the Supervising Architect's Office* (New York: Oxford University Press, 2000), 112. About the machinations behind building a new county courthouse in North Carolina during the Gilded Age, see Wayne Durrill, "A Tale of Two Courthouses: Civic Space, Political Power, and Capitalist Development in a New South Community, 1843–1940," *Journal of Social History* 35 (2002): 659–81. The Union County courthouse Durrill studied was not built of local materials, but many county courthouses were.

38. Jenni Parrish, "Litigating Time in America at the Turn of the Twentieth Century," *Akron Law Review* 36 (2002): 1–47, quotation 6. This article examines sixteen state appellate cases and one federal court decision issued between 1889 and 1924.

39. Todd Rakoff, *A Time for Every Purpose: Law and the Balance of Life* (Cambridge: Harvard University Press, 2002). In *A Moment's Notice: Time Politics across Cultures*, the anthropologist Carol Greenhouse sets "time, life, and the law" in the context of social time (175).

40. "Firemen Level Jewelers' Clocks," *Milwaukee Sentinel*, March 5, 1908. About this incident, see Lee H. Davis, "The Night the Clocks Were Torn Down," *NB* 34 (August 1992): 411–17.

41. "The Mayor and the Clocks: The Jewelers' Side of It," *Milwaukee Sentinel*, March 6, 1908, emphasis (boldfaced capital letters) in original.

42. Mayor Becker quoted in "Jewelers Cannot Recover from City," *Milwaukee Sentinel*, March 6, 1908. The full quotation reads: "Another thing about this sign business. If the city allowed the clock to remain on the streets what would prevent a saloon keeper or any tradesman from putting up a clock. A saloon could have a clock with a mug of beer or a pretzel painted on its face and the first thing the principal streets of the city would be lined with clocks."

43. "Mayor and the Clocks," emphasis (boldfaced capital letters) in original. In the midst of the controversy, the city attorney, who was a candidate for mayor, rendered the opinion that while regrettable, the "removal of the street clocks was legal." Nevertheless, other attorneys sued the city on behalf of the jewelers. "Jewelers Cannot Recover from City."

44. "Orders Jewelers to Remove Debris," *Milwaukee Sentinel*, March 7, 1908; "Jewelers Cannot Recover from City."

45. Melvin G. Holli and Peter d'Alroy Jones, eds., *Biographical Dictionary of American Mayors, 1820–1980: Big City Mayors* (Connecticut: Greenwood Press, 1981), 22–23.

46. Spokesman Archie Tegtmeyer quoted in "Jewelers Cannot Recover from City"; other arguments in support of clocks appeared in "No Decision on Clocks," *Milwaukee Sentinel*, March 10, 1908.

47. "Give Clocks to City," *Milwaukee Sentinel*, March 12, 1908.

48. Milwaukee jeweler George Durner placed an order with Seth Thomas December 8,

1908, for a two-dial electrically illuminated post clock engraved "Standard Time" and "Seth Thomas" (book I, 59 [1908], ST). Leonard Van Ess requested a post clock similar to Durner's except that he requested his name on the dial in addition to "Standard Time" (book I, 71 [1909], ST). Edwin F. Rohn ordered exactly the same post clock as Durner's (book I, 79 [1909], ST).

49. Craig, *Federal Presence*, 91.

50. Seventy-four percent of the federal government's E. Howard orders in my sample (191 out of 259) were for a total of 1,203 No. 70 clocks, which were cased wall clocks meant to hang inside buildings. The US Lighthouse Department placed the earliest orders for the No. 70 clock, the first in 1878 (box 8, 3:111 [1878], HC) and then another 36 two years later (box 8, 4:136 [1880], HC). Through the 1880s, other federal agencies installed No. 70 clocks; and in the 1890s, the bulk of the orders for them were placed. For instance, the new public building in Jackson, Mississippi, installed 11 No. 70 clocks (box 10, 3:117 [1885], HC); the Chattanooga, Tennessee, federal courthouse and post office requested 19 No. 70 clocks (box 9, 3:36 [1893], HC); the new Detroit post office ordered 42 No. 70 clocks (box 10, 2:37 [1897], HC); the Racine, Wisconsin, customs house ordered 8 No. 70 clocks (box 10, 2:112 [1898], HC).

51. See 1883 orders placed after November 18 (box 10, 3:54, 57 [1883], HC); 1884 orders (box 10, 3:59, 60, 68, 81, 83 [1884], HC); 1885 orders (box 10, 3:105, 107, 114, 115, 117, 118, 127, 130, 193 [1885], HC).

52. St. Louis (box 10, 3: 68 [1884], HC); Topeka (box 10, 3: 59 [1884], HC); Des Moines (box 9, 1:146 [1888], HC); and Council Bluffs (box 9, 1:238 [1889], HC); Craig, *Federal Presence*, 198.

53. Lee, *Architects to the Nation*, 6, 9, 111–56. In 1889, the Office of the Supervising Architect, which oversaw many federal buildings, had an inventory of 399 buildings. A little more than a decade later the number had nearly tripled to 1,126 (Craig, *Federal Presence*, 212).

54. E. Howard monopolized the market for government clocks, perhaps because of the company's reputation for manufacturing precision instruments. It fulfilled many orders from US courthouses, post offices, and other federal agencies. Between 1883 and 1898, it received sixty-six orders for interior and exterior clocks for customs houses. Only one customs house, in Nogales, Arizona, ordered a Seth Thomas clock (book E, 50 [1892], ST). Government orders from Seth Thomas were primarily for buildings associated with the military. Hour striking clocks were requested for Fortress Monroe (book B, 283 [1880], ST), San Antonio's Depot Tower (book B, 379 [1882], ST), and Fort Snelling (book B, 401 [1882], ST). National homes for disabled veterans in Santa Monica and Milwaukee ordered clock movements meant to operate hands on one exterior dial (book E, 11 [1891] and book I, 91 [1909], ST). The Sixty-Fifth Regimental Armory in Buffalo installed a two-dial tower clock (book H, 174 [1905], ST). In 1906, the Naval Academy in Annapolis requested that its Hotchkiss tower clock that operated four eight-foot dials and had a striking mechanism that rang "ship's bells as done on board" from the early 1870s be overhauled (book H, 208–9, ST). The few remaining federal orders from Seth Thomas were for exterior clocks for courthouses, post offices, and customs houses (book C, 16 [1884]; book D, 55 and 60 [1889]; book E, 50 [1892]; book H, 262 [1907]; and book I, 138 [1909], ST). Seth Thomas tower clocks were also installed at the US Army post at Fort Riley and at the National Soldiers Home in Virginia (*Tower Clocks: Manufactured by Seth Thomas Clock Co.* [1929], box 3, Warshaw-WC).

55. All the federal clocks from E. Howard's inventory in the Far West were No. 70

clocks. They could be found in the customs house in Sitka, Alaska (box 10, 2:109 [1898], HC); Sacramento's post office (box 9, 3:184 [1894], HC); the San Francisco Customs House and Appraiser's Building (box 10, 2:108–9 [1898], HC); the post office in Pueblo, Colorado (box 10, 2:84 [1898], HC); the US courthouses and post offices in Santa Fe, New Mexico (box 9, 1:280 [1889], HC), and Carson City, Nevada (box 9, 2:237 [1890], HC); two customs houses in Oregon, one in Astoria (box 9, 3:164 [1893], HC), the other in Portland (box 10, 2:108 [1898], HC); and the customs house in Port Townsend, Washington (box 9, 3:137 [1893] and box 10, 2:108 [1898], HC). About Portland's customs house and post office (1869–75), "the first permanent federal building in the Northwest," see Craig, *Federal Presence*, 123. Period photographs confirm the absence of a tower clock. A time ball was installed on top of its octagonal tower in 1885.

56. *BE*, February 1, 1898, 4.

57. Quoted in Craig, *Federal Presence*, 197. W. J. Edbrooke, the supervising architect for the Treasury Department, oversaw the design and erection of the Washington, DC, post office. Its construction began in 1891 and was finished in 1899.

58. W. Cornell Appleton, "A Government Skyscraper," *Outlook* 106 (1914): 190–91.

59. "Preliminary Drawing Upper Portion of Tower" and "Four Elevations of U.S. Custom House, Boston," Peabody and Stearns, US Custom House Plans, PS/MA.162, BPL. Preliminary drawings show only one eagle on each side of the tower, but the working drawings show two eagles beneath each clock face. Details about the clock vary from source to source. I took as authoritative Astrid C. Donnellan, "Boston's Custom House Tower Clock: A Landmark Restored," *NB* 35 (October 1993): 570–71, because she restored the timepiece.

60. "Time Ball Reports," entry 64, USHO-Records; Boston branch hydrographic office to USHO, dated July 9, 1886, entry 64, USHO-Records. The 1886 letter from Boston's hydrographic officer includes the observation about hundreds of people setting their watches by the time ball, as well as an account of interviews with the Chamber of Commerce and ship captains.

61. William Bond and Son, entry of April 8, 1885, Daybooks 1857–1903, box 15, Bond Records.

62. Winslow Upton, "Information Relative to the Construction and Maintenance of Time Balls," *Professional Papers of the Signal Service* 5 (Washington, DC: GPO, 1881), 11.

63. Two other kinds of balls were possible: one was made out of thin sheets of metal, so "that the visual effect of a solid ball is secured," and the other utilized rolled-plate copper, about one-eight of an inch thick. The former ball weighed considerably less than the latter, of which Boston's 400-pound time ball was the only existing example (Upton, "Construction and Maintenance of Time Balls," 11).

Most time balls were black; the first time ball on the US Naval Observatory was made of black India rubber. The Boston ball was copper, and some others were painted red. A few proposals bemoaned the ball's poor visibility on cloudy days. For instance, the writer of a 1908 report from the Portland, Oregon, branch hydrographic office complained: "I have approached several navigating officers upon the subject of painting the time-ball itself a different color. One that will make it a more noticeable and distinct object. It is now black. The tank and stand are also black. For half of the year the sky in this locality is heavily overcast and rain falls quite steadily. With the ball painted bright red with a wide white horizontal stripe around its equator it would offer a much better object from which to make observations. The officers all thought this an improvement and I am of the same opinion"

(July 23, 1908, USHO-Correspondence). On the other hand, an officer at the Chicago branch hydrographic office wanted to paint the red ball he was in charge of black in the effort to improve its visibility in cloudy weather (July 12, 1909, USHO-Correspondence).

64. About the time ball apparatus and how it worked, see "Time Balls," chart, 19 October 1886, Letters Received, entry 8, box 4, USNO-Records. Reports from the US branch hydrographic offices between 1885 and 1925 provide various descriptions of balls, most of which were variations on the covered frame type. See "Time Ball Reports," entry 64, USHO-Records, and USHO-Correspondence.

65. Monthly time ball reports from branch hydrographic offices that maintained time balls detail the numerous incidents that prevented the dropping of the ball between 1885 and 1927. See USHO-Correspondence and "Time Ball Reports," entry 64, USHO-Records. The USNO's records contain reports from time balls that it was responsible for, primarily the ball at the Mare Island Naval Station and the ball at Torpedo Station on Rhode Island's coast (USNO-Records, entries 5–13).

66. Quotation from Portland, Oregon, branch hydrographic office to USHO, 18 February 1911, USHO-Correspondence. In this letter the officer goes to great pains to make it clear that he "can drop the ball EXACTLY at NOON" (emphasis in the original).

67. Upton, "Construction and Maintenance of Time Balls," 11. The time ball reports often note when the press was informed of an error, and whether the officer sought through visual signals, such as lowering the ball slowly, to warn the public concerning the signals' unreliability ("Time Ball Reports," entry 64, USHO-Records, and USHO-Correspondence).

68. About the efforts of the navy and the signal service to enact favorable time ball legislation, see Ian R. Bartky, *Selling the True Time: Nineteenth-Century Timekeeping in America* (Stanford: Stanford University Press, 2000), 120–36, and Michael O'Malley, *Keeping Watch: A History of American Time* (Washington, DC: Smithsonian Institution Press, 1990), 93–94. Bartky and O'Malley's accounts differ. For facts, Bartky's should be consulted; for context and analysis, O'Malley's is best. The first piece of time ball legislation was HR 3769 (1880); in the next session of congress HR 594 (1881) was introduced. See also US Congress, House of Representatives, Committee on Commerce, 47th Congress, 1st sess., "Meridian Time and Time Balls on Custom-Houses," 1882 (HR 681). About the signal service's plans, see "Conditions on Which the Chief Signal Officer Co-Operates With Others in the Maintenance of a Public Standard Time Ball" and "Stations at Which the Signal Service Can, At Present, Maintain Time Balls," memorandums each dated 1881 in box 2, folder 4, Allen Papers. As it became obvious that uniform and correct time would have to be disseminated across the nation for the worthwhile comparison of geophysical observations, the signal service became more and more frustrated with the piecemeal, haphazard manner of timekeeping, thus its interest in disseminating the "true time."

69. In 1881 it was estimated that the simplest time ball apparatus would cost between $100 and $150; a typical time ball cost $600, the most expensive ones $2,000. Regular time signals (or the instruments to obtain the correct time) would cost considerably more than the time ball (Upton, "Construction and Maintenance of Time Balls"). A 1914 memo gives the cost of Boston's time ball and its erection in 1903 as being $1,481; Philadelphia's in 1904 as being $1,410; Baltimore's in 1905 as being $493; and the ball in Portland, Oregon, in 1906 as being $260 ("Cost of Time Balls," USHO-Correspondence).

70. A 1885 letter from the navy's chief of the Bureau of Navigation instructed the USNO superintendent: "A time-ball having been placed on the central pavilion of the Navy department building you will be pleased to discontinue dropping one from the US Naval

Observatory" (September 28, 1885, Letters Received, entry 8, box 2, USNO-Records). Gardner's system for releasing the ball was used. The technology appeared considerably more sophisticated than the one at the US Naval Observatory, which relied on a gestural or oral time signal and the manual release of the ball.

71. Unsigned letter, entry 13, vol. 5, USNO-Records. Notations in two different hands at the beginning and end of the letter suggest the author could have been "W. T. Sampson," or "Lieutenant Casey." It may be that Sampson was the letter's author; he was the USNO superintendent for a few months in 1882 and attended the Washington Meridian Conference in 1884.

72. "Time Balls," chart, October 19, 1886, Letters Received, entry 8, box 4, USNO-Records; "Programme of Work to be Pursued at the Naval Observatory, Washington, during the Year 1887," entry 8, box 5, USNO-Records. The 1886 time balls chart is useful in that it identifies the exact location of the time ball, its geographical position, its dimensions, how far the ball drops, and how far it is above sea level and ground level.

73. "Time Ball," *New Orleans Times Democrat*, February 24, 1885.

74. S. Waterhouse, "Suggestions as to the Time Ball," letter to the editor, *New Orleans Times Democrat*, February 21, 1885.

75. W. P. Ray, New Orleans branch hydrographic office, to Charles E. Black, president of New Orleans Cotton Exchange, June 8, 1885, and to USHO, June 9 and June 25, 1885, "Time Ball Reports," entry 64, USHO-Records.

76. W. P. Ray to USHO, June 25 and June 26, 1885, "Time Ball Reports," entry 64, USHO-Records.

77. President of the New Orleans Maritime Association to USHO, June 19, 1886, "Time Ball Reports," entry 64, USHO-Records; New Orleans branch hydrographic office to USHO, September 21, 1909, USHO-Correspondence. Extant time ball reports from New Orleans span the periods between October 1885 and December 1887 and February and July 1888. After July 1888, it appears that time ball reports were no longer kept.

78. Chicago and Cleveland's service began in 1899, Sault Saint Marie's in 1900, Buffalo and Galveston's in 1901, Duluth's in 1902, Norfolk's in 1904, and Key West's in 1907. See from USHO-Correspondence, memo to USNO Superintendent Jayne, August 12, 1913; Key West branch hydrographic office to USHO, August 22, 1907; "Asiatic Fleet General Order No. 19," January 6, 1908.

79. Portland, Oregon, branch hydrographic office, December 27, 1909, USHO-Correspondence.

80. See the files on Philadelphia, Baltimore, and San Francisco in USHO-Correspondence. See also J. C. Burnett, "'Tis Noon When the Time Ball Drops on the Fairmont Roof," *San Francisco Sunday Call*, May 23, 1909.

81. The requests for time balls can all be found under each city's name in USHO-Correspondence. Charles C. Carlin, my great-great-grandfather, was the Virginia congressman who requested a time ball for Alexandria.

82. The gale of September 20, 1909, blew away the New Orleans' time ball (New Orleans branch hydrographic office to USHO, September 21, 1909, USHO-Correspondence).

83. See letters, memos, and maps in New Orleans file, USHO-Correspondence. The last letter on the subject is dated February 17, 1913.

84. New York City branch hydrographic office to USHO, 6 April 1911, USHO-Correspondence.

85. "Give Lighthouse for Titanic's Dead," *NYT*, April 16, 1913; "Time Ball on Titanic Memorial to Drop for First Time To-Day," undated news clipping, Arthur Collection.

86. Letters of May 11 and May 24, 1920, USHO-Correspondence.

87. Stephens, "Most Reliable Time," 9–11; O'Malley, *Keeping Watch*, 109, 121, 133; Bartky, *Selling the True Time*, 153–54. About daylight saving time, see O'Malley, *Keeping Watch*, 256–96; Barky, *Selling the True Time*, 2, 4, 207; Ian R. Bartky, *One Time Fits All: The Campaigns for Global Uniformity* (Stanford: Stanford University Press, 2007), 161–210; David Prerau, *Seize the Daylight: The Curious and Contentious Story of Daylight Saving Time* (New York: Thunder's Mouth Press, 2005); James Kenneally, "'An Hour of Light for an Hour of Night': Daylight Saving in Massachusetts," *New England Quarterly* 78 (September 2005): 440–47; Adam Barrows, *The Cosmic Time of Empire: Modern Britain and World Literature* (Berkeley: University of California Press, 2011).

88. George Rockwell Putnam, *Lighthouses of the United States* (Boston: Houghton Mifflin, 1917), 52. Lighthouses were, according to historian John Stilgoe, the "first structures over which local communities had no control" (John Stilgoe, *Common Landscape of America, 1580–1845* [New Haven: Yale University Press, 1983], 110). As he explains, "the builders and keepers answered to the new national government" (110). These beacons demonstrated "the new strength of nationalism over localism, the new power of new government" (110). Stilgoe's interpretation of the earliest federal program of lighthouse construction could apply as well to the vogue for time balls: "The federal government may well have understood, if only vaguely, that the towers symbolized its strength and that every man, woman and child who saw their massiveness might glimpse the permanency and strength of the infant republic" (111). Stilgoe compares lighthouses to rolands or columns that "order the center of European cities and recall the time when a city's greatest possession was its privilege of self government" (111).

CHAPTER SIX

1. "What Time Is It?" *NA*, May 4, 1899, clipping, Perkins Scrapbook. The complaint concluded with the quip, "Some of us are interested in the passage of time, even if the Commissioners are so indifferent to it as not to care whether the Hall is finished this year or sometime in the next century." See also Michael J. Lewis, "'Silent, Weird, Beautiful': Philadelphia City Hall," *Nineteenth Century*, v. 11 (1992): 13–21.

2. About the clocks mentioned, see box 4, 29:314 (1889), HC; box 9, 2:89, 104, 167, 195, 219 (1890), HC; box 10, 3:59 (1884), HC; book B, 358 (1882) and book E, 15 (1891), ST. See also Moses King, *Philadelphia and Notable Philadelphians* (New York: Moses King Publishers, 1902), 8. The rate of clock orders accelerated in the 1880s, which could have been due to a number of factors (such as E. Howard's and Seth Thomas's consolidation of the public clock business) in addition to increased demand for public clocks. Three times as many clocks were ordered from E. Howard and Seth Thomas in the 1880s as in the 1870s. The number of clock orders in the 1890s was 10 percent higher than in the 1880s, which is impressive considering that between 1893 and 1897 the nation underwent its most serious economic contraction to that date.

3. "Tower Clock Starts in 1899," *PT*, December 31, 1898. A pneumatic system sent impulses from the clockworks to the clock dials.

4. Time balls had been displayed at earlier expositions, including ones in New Orleans in 1884 and Cincinnati in 1888. On the opening day of the Centennial Exposition of the Ohio Valley and Central States, July 4, 1888, the wife of former president James K. Polk presided over the reception of a noon time signal sent from Nashville (telegram,

July 4, 1888, entry 15, USNO-Records). The USNO sent three time balls to the exposition (invoice of public property transferred by the US Naval Observatory to B&O Railroad, June 30, 1888, entry 9, USNO-Records). The officer in charge of the exhibit wrote: "There is a great deal of prejudice here regarding standard time, but the Mayor and several prominent citizens have expressed themselves strongly in favor of Obsy. time. I try to do all the missionary work I can and avoid newspaper controversy" (letter from A. B. Clements, August 4, 1888, entry 9, USNO-Records).

5. "Report of the State of the Naval Observatory Exhibit," May 31, 1893, and A. G. Winterhalter to USNO, May 8 and May 25, 1893, box 3, entry 11, USNO-Records; Charles F. Carpenter, "Horology," in *Report of the Committee on Awards of the World's Columbian Commission* (Washington, DC: GPO, 1901), 1:885.

6. Ibid., 891. One diarist recounts: "In the upper part of the middle section [of the Terminal Building] we saw something of great interest to us. It was the dials of twenty-four clocks, about four or five feet in diameter. They gave the time of day, or night, at twenty-four of the principal cities of the world. At the time we saw them it was just 10:30 a.m. in Chicago; the clocks gave the time in the different places" (Plooma M. Boyd, Columbian Exposition diary, 1893, Archives Center, NMAH).

7. *A Week at the Fair* (Chicago: Rand, McNally, 1893), 140, 136, 66, 224, 215; Paul R. Strain, Columbian Exposition diary, 1893, Archives Center, NMAH; "Report of the State of the Naval Observatory Exhibit"; Gary Nash, *The Liberty Bell* (New Haven: Yale University Press, 2010), 156–61.

8. Carpenter, "Horology," 886–87; Strain diary.

9. Carpenter, "Horology," 888–89, 890, 916. The Tiffany exhibit also displayed a globe clock and nineteen precious watches in gold cases. The sports represented on the Goldsmith's and Silversmith's clock were swimming, running, yachting, cycling, baseball, trotting, and jumping. Among the sporting figures was a lacrosse player. The California Museum of Photography, part of the University of California, Riverside, holds a Keystone-Mast stereograph of the clock; its identification number is KU71659. One of the many exposition diaries from 1893 describes the clock at some length (Boyd diary).

10. Robert Rydell, *All the World's a Fair: Visions of Empire at American International Expositions, 1876–1916* (Chicago: University of Chicago Press, 1984); Matthew Frye Jacobson, *Barbarian Virtues: The United States Encounters Foreign Peoples at Home and Abroad, 1876–1917* (New York: Hill and Wang, 2000). There were several timekeeping exhibits at the 1939 New York World's Fair, where the clock dial on the Pennsylvania State Building's tower was not attached to clockworks, so it permanently showed ten minutes to three as the time (Meyer Berger, "At the World's Fair Yesterday," *NYT*, July 2, 1939). The 1964 New York World's Fair, an unofficial trade fair, featured "the most dazzling display of Swiss wristwatches ever seen in America." The Swiss Pavilion, known as the Time Center, featured a master clock connected to ten clock towers spread throughout the fairgrounds (Bruce Shawkey, "Swiss Watch Industry Sparkles at 1964 New York World's Fair," *NB* 50 [April 2008]: 155–58).

11. Thomas Bartels, "The 8th Wonder of the World," *NB* 32 (February 1990): 16–24.

12. All quotations are from "A Remarkable Clock," *Manufacturer and Builder* 12 (1880): 210. This notice refers to a clock made by Felix Meier (also spelled Meyer) in Detroit over the course of the 1870s. See also H. W. Ellison, "The Wonderful Meier Clock," *NB* 32 (February 1990): 25–28.

13. Michael O'Malley and Carlene Stephens, "Clockwork History: Monumental Clocks and the Depiction of the American Past," *NB* 32 (February 1990): 3–15. See also Chris

Bailey, "Mr. Wegman's American History Clock," *NB* 32 (February 1990): 29; Ellison, "Meier Clock"; Bartels, "8th Wonder of the World."

14. Donald Saff, "Smallwood's 'Masterpiece,'" *NB* 50 (February 2008): 23–35.

15. Stuart Leuthner, "The Apostolic Clock: The Restoration of a Timekeeping Masterpiece," *NB* 53 (April 2011): 150–54. Hughes' clock is now part of the Erie County Historical Society's collections. Another apostolic clock from the period is the Fiester clock, which was first displayed at the Penn Street Fair in Reading, Pennsylvania, in 1878, and now is part of the Hershey Museum's collections. James D. McMahon Jr., "The Ninth Wonder of the Age: The John Fiester Monumental Apostolic Clock Unveiled," *NB* 45 (August 2003): 425–35.

16. Entry for "Great Historical Clock of America," NMAH-SI, http://collections.si.edu/search/results.jsp?q=record_ID:nmah_852074 (accessed January 24, 2012). For an account of the clock in Sydney, Australia, see "Historical American Clock," *Sydney Morning Herald*, August 2, 1893.

17. CDN, February 21, 1908; Barbara Ching, "'This World of Ours': The Bily Clocks and Cosmopolitan Regionalism, 1913–1948," *Annals of Iowa* 68 (Spring 2009): 111–36; Lee H. Davis, "Time in Miniature," *NB* 52 (June 2010): 265–72; O'Malley and Stephens, "Clockwork History," 4.

18. The engraving was published by the Treasury Department's Supervising Architect's Office in 1888. It is reproduced in Lois A. Craig, *The Federal Presence: Architecture, Politics, and Symbols in U.S. Government Building* (Cambridge: MIT Press, 1984).

19. Michael Adas, *Machines as the Measure of Men: Science, Technology, and Ideologies of Western Dominance* (Ithaca: Cornell University Press, 1989). The anthropologist Carol J. Greenhouse has deftly dissected ethnographic bias concerning the temporal sense of "primitive peoples" in *A Moment's Notice: Time Politics across Cultures*, (Ithaca: Cornell University Press, 1996). See also Johannes Fabian, *Time and the Other: How Anthropology Makes Its Object* (New York: Columbia University Press, 1983).

20. The best-known instance of this is Edward S. Curtis's photograph *In a Piegan Lodge* (1910). The touched and untouched versions are digitized at LC-PPD. See Christopher Lyman, *The Vanishing Race and Other Illusions: Photographs of Indians by Edward S. Curtis* (New York: Pantheon Books, 1982), 86, 106; Gerald Vizenor, "Edward Curtis: Pictorialist and Ethnographic Adventurist," (2000) on the American Memory website of the Library of Congress, http://memory.loc.gov/ammem/award98/ienhtml/essay3.html (accessed March 6, 2012).

21. "Clock Starts with New Year: Monster Timepiece in City Hall Will Begin Its Work to-night," *PP*, December 31, 1898; "The Old Clock Must Come Down," *NA*, July 21, 1898, clipping, Public Property Scrapbooks; T. Mellon Rogers, diary, February–July 1898, typescript, entries of April 13 and July 15, 1898, IHA. The Daughters of the American Revolution, who had charge of the building, hired Rogers to develop plans for the restoration of Independence Hall.

22. *PL*, July 1, 1898. See also Frank M. Riter, statement about restoration of Independence Hall, undated document in *Documents Relating to the Physical History of Independence Hall*, reproduced in *Philadelphia Times Sunday Special*, July 3, 1898.

23. "Proposed Removal of Independence Hall Clock," *PL*, July 22, 1898, and "The Old Clock Must Come Down," *NA*, July 21, 1898.

24. Quotations from "The Clock Must Go," *PT*, July 21, 1898, and "Change of Clock Dials," *PL*, August 1898, clipping, Campbell Scrapbook Collection, 85:10–11, HSP.

25. "Rededication of Independence Hall," *PL*, October 29, 1898.

26. Curator of Independence Hall to chief of Bureau of City Property, January 24, 1918, folder 17, box 24, ser. 1, subser. A, BCP-IH.

27. City of Philadelphia, September 6, 1898, contract with Johnson Temperature Regulating Company for Installment of Tower Clock, 160.26, PCA. See also "Clock Committee Report," typewritten, April 30, 1898, included in Committee, Metal Work of Tower, Minutes, 160.9, PCA; Commissioners for the Erection of Public Buildings, *Guide to the City Hall, Philadelphia*, 1899, 35, City Guides Collection, PCA; "Tower Clock Starts in 1899."

Warren S. Johnson, who bid $27,900, received the contract for the clock ("The City Hall Tower Clock," *Telegraph*, June 23, 1898; Warren S. Johnson, "The Philadelphia City Hall Clock," *Journal of the Franklin Institute* 151 [February 1901]: 81–107). Milwaukee's city hall clock also relied on Johnson's pneumatic system.

28. *Turner's Guide to and Description of Philadelphia's New City Hall* (Philadelphia, 1892), 30, City Guides Collection, PCA. See also Johnson, "Philadelphia City Hall Clock," esp. 84.

29. The bronze statues of the allegorical group, completed between 1894 and 1896, include an Indian woman and child; an Indian chief and dog; a Swede woman, child, and lamb; and a Swede man and boy (Francis Schurmann, *Monograph on Design and Construction of City Hall Tower*, typescript, 1898, 7–9, City Guides Collection, PCA). On the decorative use of eagles in Philadelphia, see Robert F. Looney, ed., *Old Philadelphia in Early Photographs, 1839–1914* (New York: Courier Dover, 1976). "Ornamental eagles were frequently used above store fronts, hotel cornices and clock towers" (87).

30. Lewis, "Philadelphia City Hall," 13. "At three minutes before nine o'clock each evening, the corona of arc lamps encircling the Tower at the upper platform is extinguished, and again lighted at precisely nine o'clock" (*Guide to the City Hall, Philadelphia*, 1899, 36, City Guides Collection, PCA).

31. The Tower Clock," *PTel*, September 9, 1898; untitled and undated news clipping, Perkins Scrapbook; "Progress of City Hall Clock," *PR*, November 4, 1898; "The Tower Clock Starts Ticking," *PTel*, December 1, 1898; invitation, December 28, 1898, Perkins Scrapbook.

32. "City Hall Clock Starts: Monster Timepiece in Penn's Pedestal Begins Business with the New Year," *PP*, January 1, 1899.

33. On December 31, 1904, the New York Times Company held a New Year's celebration with fireworks that drew two hundred thousand people to the newly named Times Square, where their new building, the Times Tower would publish its first issue on January 2, 1905. Because the company could not get a permit for fireworks to celebrate the arrival of 1906, it lowered an illuminated glass ball weighing four hundred pounds from the flagpole at midnight. Two years later to welcome 1908, publisher Adolph S. Ochs commissioned a New Year's Eve Time Ball, which dropped at midnight December 31, 1907. Even when the company moved in 1913, the time ball continued to drop on New Year's Eve in Times Square. About this, see Scott Huler, "Meanwhile: A Brief History of Time Balls," *International Herald Tribune*, December 31, 2004. Huler astutely observes that on New Year's Eve when the ball drops, "Americans will, together, do something that has otherwise become an almost entirely independent and private activity: they will tell the time." He calls it "the greatest single moment of public timekeeping in the world."

34. In late 1893, the reverend in charge of New York's Trinity Church decided against ringing in the New Year with the church's bells ("Trinity's Chimes to be Silent," *NYT*, December 31, 1893). The chimes rang again the next year, but by 1900 it was reported

they could not be heard ("Every Peal of the Old Church Chimes Drowned in the Bedlam of Discordant Notes," *NYT*, January 1, 1900).

35. Quotations from "Big Tower Clock Begins the New Year," *NA*, January 2, 1899, and "City Hall Clock Starts," *PL*, December 3, 1898. There were plenty more observations in the same vein: "The massive hands of the municipal clock will begin their onward motion, which, barring accidents, will cease only with the downfall of its lofty pedestal" ("Tower Clock Starts in 1899").

36. "Storm Stops Clock Hands," *PP*, January 2, 1899. See also "City Hall Clock Starts."

37. "City Hall Tower in Darkness," *PR*, May 8, 1899, clipping, Perkins Scrapbook. See also "No Liquor There," *PTel*, May 2, 1899.

38. "City Hall Clock Starts." Reports about the clock's failure to keep time include the following clippings found in Perkins Scrapbook: "City Hall's Queer Antics," *PT*, May 1, 1899; "Town Clock Torpid," *PP*, May 1, 1899; "What Time Is It?," *NA*, May 4, 1899. See also the clockmaker Warren S. Johnson's refutation of the charges that his clock was unreliable: Johnson, "Philadelphia City Hall Clock."

39. E. Howard Clock Company, *Regulators, Office and Bank Clocks*, [1901], 19, box 3, Warshaw-WC.

40. Nick Yablon, *Untimely Ruins: An Archaeology of American Urban Modernity, 1819–1919* (Chicago: University of Chicago Press, 2009), 191–242. Yablon insightfully discusses the vast amount of disaster photographs taken after the earthquake in a chapter titled "'Plagued by Their Own Inventions': Reframing the Technological Ruins of San Francisco, 1906–1909."

41. James S. Tyler, "When the Old Ferry Clock Came Down," *San Francisco Call*, June 17, 1906.

42. "Ferry Clock Helps Herald New Year," *San Francisco Call*, January 1, 1907. The newspaper reported that it was the first time since the earthquake that the dials were illuminated so that its hands could record "the shifting hours of the dying old year."

43. Michael Tavel Clarke, *These Days of Large Things: The Culture of Size in America, 1865–1930* (Ann Arbor: University of Michigan Press, 2007). Clarke explores the discourse of size as it related to physical bodies, cities, and buildings. He does not discuss clocks.

44. "The Largest Clock in the World," *Scientific American* 98 (May 23, 1908): 375; "Twenty Foot Hand for Tower Clock," unidentified news clipping (1908), Arthur Collection; Colgate Clock advertisements in *New-York Tribune*, June 5, 1908, and *Evening World*, June 1, 1908. See also "Biggest Clock Running," *New-York Tribune*, May 26, 1908; "Gossip of the Commuters," *New-York Tribune*, May 3, 1909.

45. Yablon, *Untimely Ruins*, 279–84; Nick Yablon, "The Metropolitan Life in Ruins: Architectural and Fictional Speculations in New York, 1909–19," *American Quarterly* 56 (June 2004): 308–47.

46. "Two New Hands Point Out Time for Broadway," *New York Herald Tribune*, August 10, 1935, clipping, Arthur Collection. Completed in 1926 for the price of $40,000, the Paramount's dial and the stationary time ball each depended on fifty thousand watts of electric power for illumination.

47. Ibid.

48. Yablon, *Untimely Ruins*, 246; see also 280, 282–83

49. *Evening Telegraph*, March 15, 1900; "Pierie Defends His Dummy Clock," *PL*, April 10, 1900; untitled article, *PT*, April 13, 1900.

50. Quoted in untitled article in *PL*, April 10, 1900.

51. Ibid. It had been rumored, based on an interview with the chief clerk in the Bureau of Public Property, that before the previous chief of public property had retired from office, it had been decided to abandon the project of a running clock, because it was considered unnecessary and expensive. The rumor had it that for these reasons the office decided to install a dummy clock instead.

52. "Public Refused to be Fooled by Spurious Clock," unidentified news clipping, April 13, 1900, Public Property Scrapbooks; "Dummy Clock to be Removed," undated clipping [ca. April 1900], Public Property Scrapbooks.

53. "Old Congress Hall," *PL*, March 1, 1901; "Public Refused to be Fooled"; "War of the Two Clocks; or, a Brief Story of How a Restoration Convulsed a City," *PP*, April 22, 1900.

54. Quotations from the *Dodge-City Globe-Republican*, August 18, 1898. In 1888, New York's *Evening World* printed a piece titled "A Silly Story." It read: "The New York *Sun* has started a pretty little story about why the hands on all dummy clocks used for jewellers' and watchmakers' signs point to eighteen minutes past eight. That newspaper says that the principal maker of these dummy signs in New York was working on a big clock on the evening of April 14, 1865, when in rushed the jeweller who had given him the order. He was just painting the hands as the customer said: 'Point those hands at the hour Lincoln was shot, that the deed may never be forgotten.' So the painter put the hands at eighteen minutes after eight. Unfortunately for the *Sun's* romance, Mr. Lincoln was shot at exactly half past ten o'clock. While it may be true that all dummy clocks have their hands pointed at eighteen minutes past eight, the reason assigned will have to be sought further" (November 21, 1888).

With this piece, the *Evening World* was refuting an article in the *Sun* titled "One Puzzle Solved: Why All Painted Clocks Point to Eighteen Minutes after Eight," which was widely reprinted around the nation for the next quarter of a century. See *Washington Evening Post*, December 20, 1888; *Columbus (NE) Journal*, January 9, 1889; *Washington Critic*, February 9, 1891; *Omaha Daily Bee*, September 21, 1891; *Washington Morning Times*, March 4, 1897; *Kansas City Journal*, June 17, 1897; "Jewelers' Dummie Clocks," *San Angelo Press*, April 2, 1902; "Diary of Father Time," *Chicago Day Book*, February 17, 1914. Curiously, two years after refuting the *Sun's* "silly story," the *Evening World* itself ran the story without commentary (September 24, 1890). For alternative stories about the placement of clock hands on dummy clocks, see "Dummy Clocks of London Jewelers," *Mt. Sterling (KY) Advocate*, October 17, 1899; "A Curious Fact," *San Juan Islander* (Friday Harbor, WA), February 16, 1907.

In the meantime, in 1894 the *Sun* ran a letter from a New York City jeweler stating that painting hands to 8:18 created enough room on the dials for the jeweler to feature his name and wares ("The Hands of a Clock," *Sun*, June 3, 1894). The *Saint Paul Globe* published a nice summary and refutation of the legend titled "Lincoln's Death and the Dummy Clocks" (October 14, 1901).

Cheryl A. Wells found that when Abraham Lincoln died, "clocks were either stopped or interrupted" and "bells set to toll the time stopped" (*Civil War Time: Temporality and Identity in America, 1861–1865* [Athens: University of Georgia Press, 2005], 114).

55. "New Hour of Fate," *Minneapolis Journal*, September 19, 1901; "Clock Hands to be Changed," reprinted from the *Denver Post* in *Anadarko (OK) Daily Democrat*, November 7, 1901; "Clocks to Indicate Assassination Hour," *San Francisco Call*, December 7, 1901. See also *Clarke Courier* (Berryville, VA), December 18, 1901.

56. For specified memorial clock orders, see the following examples in the E. Howard

Collection: box 10, 3:46, 49 (1883); box 9, 3:294, 297 (1894); box 10, 2:10, 70 (1897); box 11, 1:180 (1905).

57. Greenhouse, *Moment's Notice*, 19–48.

58. Frank T. Lee, *Some Practical Lessons in Life, Suggested by the New Memorial Tower Clock* (Whitewater, WI: Register Steam Printing House, 1887), 3. A clock tower constructed in 1840s in St. Louis had engraved on its frame "All the hours quickly pass. May the last be happy for you" (Frederick M. Shelley, *Early American Tower Clocks: Surviving American Tower Clocks from 1726 to 1870* [Columbia, PA: NAWCC, 1999], 204).

59. See Lee, *Some Practical Lessons*, 5–7, 11, 9.

60. Michael O'Malley, *Keeping Watch: A History of American Time* (Washington, DC: Smithsonian Institution Press, 1990), 150–51, 172–99. This was not an unusual opinion for the period; George Beard's *American Nervousness* publicized its broad outlines.

61. Lee, *Some Practical Lessons*, 12.

62. A one-dial tower clock and the inscribed bell were for the Krauth Memorial Library (book H, 268, ST). The inscription included the psalm numbers from which the text was quoted.

63. The Ellen Brown clock (box 9 3:109 [1894], HC); the order from Brooklyn's Greenwood Cemetery for five-foot clock dials to be mounted on a nine-foot-high tower at the cemetery's entrance gate (box 9, 3:93 [1893], HC); Long Island's Pinelawn Cemetery's order for a clockwork to run four galvanized iron four-foot dials (box 11, 1:183 [1905], HC); the Harvard clock (box 10, 2:18 [1897], HC and an undated article, *BE*, April 19, 1897). Memorial clocks were also ordered from Seth Thomas (book C, 52, 54 [1885], book D, 18–19 [1887], book H, 34 [1903], ST). Thirty-three clock orders placed between 1874 and 1909 from E. Howard and Seth Thomas specified that the clocks were meant as memorials.

64. "Address of Presentation at Louvain, July 4, 1928," in *The American Engineers' War Memorial in the Tower of the Louvain Library: The Book of Names* (NY: Bartlett Orr Press, 1928). Numerous memorial clocks were erected across Europe and the United States after World War I, though not after World War II. Sydney Algernon Logan erected a sixty-five-foot-high "Peace Tower," with a single clock dial and bell (Shelley, *Early American Tower Clocks*, 201).

65. *New-York Tribune*, February 5, 1888; "The Clock Deceived Him," *Daily Yellowstone Journal* (Miles City, MT), April 3, 1890.

66. Some clock companies specialized in the production of advertising clocks. The Baird Clock Company, for example, embossed messages like "It Saves Time" and "For all Cleaning Sapolio Is Best" on the doors of cased tin-dialed clocks. See Jerry Maltz, "Baird Clock Co.: The Clocks and the Companies that Advertised," *NB* 50 (August 2008): 418–24.

67. "Public Refused to be Fooled"; "Dummy Clock to be Removed." Interestingly, the restoration committee had intended for the clock to run. Specifications had been drawn that would have produced a running clock, and a contractor had been hired for $18,000 to do the work. Blueprint: "Restoration of the State House—Detail of Clock," approved December 10, 1898, folder 5, ser. 1, subser. D, BCP-IH. This blueprint shows a clock face with the hands set to 1:50. Contract and specifications: "Supplemental Contract for Restoration of Independence Hall, with Specifications Attached," March 28, 1899, reproduced in *Documents Relating to the Physical History of Independence Hall*, vol. 2, IHA.

68. Posted on the website of "Witness Watch," https://witnesswatch.wordpress.com/2009/11/15/222hundreddollarbillclock/ (accessed June 18, 2012). See also the theories

about the significance of the time shown on the clock on the website FunTrivia, http://www.funtrivia.com/askft/Question34654.html (accessed June 18, 2012).

69. "Big Tower Clock Begins the New Year," *NA*, January 2, 1899, clipping, Perkins Scrapbook.

70. "The State House Dials," *PP*, July 21, 1898, clipping, Public Property Scrapbooks.

71. Leonard Beale to superintendent of buildings, February 26, 1920, folder 312, box 19, ser. 1, subser. A, BCP-IH; Frazier, "Centennial Clock," 52.

72. All of the following items are from box 19, ser. 1, subser. A, BCP-IH: letter to mayor of Philadelphia, January 19, 1917, folder 27; "Toll Old State House Bell for Mrs. Wilson," news clipping, August 11, 1914, folder 20; memo from chief of Bureau of City Property, February 4, 1924, folder 33; memo from superintendent of city hall, January 6, 1933, folder 34.

EPILOGUE

1. "City Hall Afire," *BE*, February 26, 1895. The city hall, built between 1845 and 1849, was the oldest extant building in Brooklyn. See William Lebovich, *America's City Halls* (Washington, DC: Preservation Press, 1984), 49.

2. *BE*, March 10, 1895, 10.

3. "The City Hall Clock," *BE*, April 21, 1895; "To Improve the City Hall," *BE*, May 26, 1895.

4. "City Hall Tower Confab," *BE*, June 6, 1895, 4. Proposals for the repairs are detailed in "For the City Hall Tower," *BE*, June 3, 1895, 2.

5. "War at City Hall," *BE*, June 7, 1895, 6. An advisory committee to Brooklyn's mayor, commissioner of public works, and commissioner of public buildings opted for the tower, largely because it "would be alike visible and effective from every point, near or distant, high or low" (*BE*, June 18, 1895, 1). See also "City Hall Belfry," *BE*, June 19, 1895, 14.

6. "The City Hall Tower," *BE*, October 2, 1896.

7. "No Clocks for Bingham," unidentified news clipping, 1908, Arthur Collection.

8. A note dated 1892 penciled in beside an 1885 order for a No. 16B hour striking clock placed by local jeweler named A. H. Rhoades for an unidentified church in Pottstown, Pennsylvania, sums up some of the impediments facing timekeepers: "Marion Bradley [clock repairman] says clock is up in shape, but is not well protected. Rhoades tried to be cheap. Clock in charge of Sexton of Church" (book C, 77 [1885, 1892], ST).

9. James H. Collins, "Electric 'Time Ball': South American System of Adjusting Watches Suggested for This Country," *NYT*, July 3, 1921. Collins visited South America during the previous year, where he was impressed with the utility of flashing the time. He described how the Uruguayan government cut the electric current for a moment every night at eight so people could set their watches.

10. Barbara Adam, *Time* (Cambridge: Polity Press, 2004), 118–19. See Ralph E. Gould, *Standard Time throughout the World*, Circular of the National Bureau of Standards C406 (Washington, DC: GPO, 1935).

11. Michael A. Lombardi, "Time Signal Stations," *Popular Communications*, February 2006, 8–19, posted on the website of the National Institute of Standards and Technology

(NIST), Time and Frequency Division http://tf.boulder.nist.gov/general/pdf/2131.pdf (accessed March 17, 2012).

12. Adam, *Time*, 119.

13. For an example of an argument that computers change the ways humans measure and order time, see Jeremy Rifkin, *Time Wars: The Primary Conflict in Human History* (New York: Henry Holt, 1987), 191, 193.

14. Folders 30 and 31, Christine Frederick Collection (MC261), Arthur and Elizabeth Schlesinger Library on the History of Women in America (Radcliffe Institute for Advanced Study, Harvard University). See folder 31, photos 1 and 2, for pictures of men setting timepieces to radio time.

15. Much more research needs to be done about the role radio (and electronic broadcasting in general) has played in the dissemination of time over the course of the twentieth century. The testimony of "radio men"—including representatives from the Columbia Broadcasting System, the National Broadcasting Company, Pacific Northwest radio stations, and the National Association of Broadcasters—to the US Senate's Uniform Time hearings held in 1948 is fascinating. See *Uniform Time Hearing before a Subcommittee of the Committee on Interstate and Foreign Commerce, United States Senate* (Washington, DC: GPO, 1948).

16. Seth Thomas Clock Company to Francis G. DuPont, February 2, 1883, and November 15, 1885, in DuPont Papers, ser. 1, "Scientific and Astronomical Papers," box 38. Quotation about observatory in Francis G. DuPont to Standard Electric Time Company (SETC), May 26, 1890, in DuPont Clock Correspondence. More generally, see William S. Dutton, *DuPont: One Hundred and Forty Years* (New York: Charles Scribner's Sons, 1942), 124, 140, 159–60.

17. Francis G. DuPont to the SETC, March 18, 1891, and September 25, 1890, DuPont Clock Correspondence.

18. Francis G. DuPont to the SETC, October 17, 1897, DuPont Clock Correspondence.

19. Ian R. Bartky, *Selling the True Time: Nineteenth-Century Timekeeping in America* (Stanford: Stanford University Press, 2000), 167–83.

20. Companies that sold what were called "master-slave" clock systems included the Standard Electric Time Company, the Time Telegraph Company, and the Waltham Clock Company. Companies that sold electrically wound clocks (or "self-winding" clocks) included the Self Winding Clock Company, the American Self-Winding Clock Company (Chicago), the National Self-Winding Clock Company (Bristol, Connecticut), the No-Key Clock Company (West Virginia), the Trinity Electric Clock Company (Chicago), the Tiffany Never-Wind Clock Company, and the Waltham Clock Company. There is no single reference for the history of these devices or companies. See the following useful articles: J. Alan Bloore, "School Clock Systems of the Standard Electric Time Company," *NB* 53 (April 2011): 187–99; Ray Brown, "'You Never Have to Wind Them': A History of the Trinity and American Self Winding Clock Companies of Chicago," *NB* 43 (February 2011): 71–79; Leonard Brenner, "A History of George Steele Tiffany Clocks," NB 48 (February 2006): 57–65; Andrew H. Dervan, "Waltham Clock Company," *NB* 47 (April 2005): 193–207; Dervan, "Waltham Clock Company—the Electric Years," *NB* 50 (August 2008): 425–27.

21. Jim Linz, *Electrifying Time: Telechron and GE Clock, 1922–1925* (Atglen, PA: Schiffer Publishing, 2001); Ray Brown, "Time Machines: A Review of Twentieth-Century

AC Clocks," *NB* 51 (February 2009): 20–24; Bill Keller, "The Sangamo Story," *NB* 50 (August 2008): 455–56.

22. Philadelphia's William Heine (1880–1976), for instance, had five assistants in his clock shop in the 1920s. Together they serviced twelve hundred clocks weekly; most of their work was known as "regulating clocks" which included winding them and setting them to the correct time. By the 1940s, Heine had fifteen employees and had taken charge of Philadelphia's tower clocks; he redesigned the city hall clock in 1948 so it could be powered with electricity. See Bruce R. Forman, "A History of William A. Heine," *NB* 46 (February 2004): 3–8.

23. Francis G. DuPont to the SETC, January 25, 1899, DuPont Clock Correspondence.

24. Francis G. DuPont to the SETC, January 15, 1899, DuPont Clock Correspondence.

25. Jack Linahan, "Watch and Clock Collector, Henry Ford, b. 1863–d. 1947," *NB* 46 (October 2004): 579–85.

26. The preponderance of pocket watches hanging in a watch repairman's shop in 1939 suggests that people were still using pocket watches in the late 1930s ("Jeweler and watch repair man San Augustine, Texas," photography by Russell Lee, April 1939, US Farm Security Administration, LC-USF34-032933, LC-PPD).

In 1941, *Reader's Digest* declared in an article headline, "The Watch Repair Man Will Gyp You if You Don't Look Out." It came to this conclusion after field investigators found that nearly half of the 462 watch repairmen they visited "lied, overcharged, gave phony diagnoses, or suggested extensive and unnecessary repairs." It must have been nearly a truism that this was the case if *Reader's Digest* commissioned the study. Indeed, it was part of a larger series by the consumer investigators Roger William Riis and John Patric about automobile, radio, and appliance repairmen who cheated customers. See Roger William Riis and John Patric, *Repairmen Will Get You If You Don't Watch Out* (New York: Doubleday, Doran, 1942); reprinted as *Repair Men May Gyp You* (Kingsport, TN: Kingsport Press, 1951). In the 1950s, two economists who looked into the pricing of watch repairs came to a similar conclusion. See Fred L. Strodtbeck and Marvin B. Sussman, "Of Time, the City, and the 'One-Year Guaranty': The Relations between Watch Owners and Repairers," *American Journal of Sociology* 61 (May 1956): 602–9.

27. About wristwatches, see Helmut Kahlert, Richard Mühe, and Gisbert L. Brunner, *Wristwatches: History of a Century's Development*, trans. Edward Force (1983; reprint, West Chester, PA: Schiffer Publishing, 1986); Carlene Stephens and Maggie Dennis, "Engineering Time: Inventing the Electronic Wristwatch," *British Journal for the History of Science* 33 (December 2000): 478–99; Bruce Shawkey, "Art Deco," *NB* 47 (December 2005): 700–703; Shawkey, "Early Waltham Watches," *NB* 51 (June 2009): 310–12; Shawkey, "Depollier Cases and Watches—a Shooting Star of the Teens and 20s," *NB* 50 (February 2008): 12–15; Shawkey, "Collectors Swoon over '20s Ladies' Masterpiece," *NB* 52 (October 2010): 537–41; Bryan Girouard and Will Roseman, "The 0-Sized Wristwatch: Hamilton's First Wristwatch for Men," *NB* 48 (April 2006): 167–73.

28. Quoted in Shawkey, "Early Waltham Watches," 311.

29. Michael C. Harrold, "An Economic Look at the American Watch Industry: The Pocket Watch Era, 1860–1930," *NB* 49 (August 2007): 425–41, esp. 425.

30. Hai Ren, "The Countdown of Time and the Practice of Everyday Life," *Rhizomes: Cultural Studies in Emerging Knowledge* 8 (Spring 2004), http://www.rhizomes.net/

issue8/ren.htm (accessed June 21, 2011); Ren, "The Merit of Time: A Genealogy of the Countdown," in *The End that Does: Art, Science, and Millennial Accomplishment*, ed. Cathy Gutierrez and Hillel Schwartz (London: Equinox, 2006); Ren, *Neoliberalism and Culture in China and Hong Kong: The Countdown of Time* (New York: Routledge, 2010).

31. Brian Stelter, "The 'Countdown Clocks' Became Part of the Story," *NYT*, April 8, 2011.

32. D. B. Sullivan, "Time and Frequency Measurement at NIST: The First 100 Years" (2001), http://tf.nist.gov/general/pdf/1485.pdf (accessed March 17, 2012); Claude Audoin and Bernard Guinot, *The Measurement of Time: Time, Frequency, and the Atomic Clock* (Cambridge: Cambridge University Press, 2001). About quartz watches, see Carlene Stephens and Maggie Dennis, "Engineering Time: Inventing the Electronic Wristwatch," *British Journal for the History of Science* 33 (December 2000): 478–99.

33. Quoted on NIST's History Measuring Time and Frequency webpage, http://www.nist.gov/pml/div688/grp40/division-history.cfm (accessed March 17, 2012).

34. Network Time Protocol and Precision Time Protocol each disseminate UTC: they are techniques developed in the mid-1980s that synchronize clocks using computer networks and systems.

35. Information about the US government's clocks and time services is available on the webpage of the NIST Time and Frequency Division, http://www.nist.gov/pml/div688/ (accessed March 17, 2012).

36. Michael Lou, "Got the Time? At Grand Central, It Has Never Been that Simple," *NYT*, July 6, 2004. Penn Station adopted a similar system for synchronizing its clocks in 2003. Thanks to Nick Yablon for sending me this article.

INDEX

Page numbers in italics refer to tables and illustrations.

146; theft of, 42, 70–71, 199n30, 212n13; uses of, 64, 78, 81–88; and women, *80*, 84–85, *86*, 213n22; wrist, 9, 179, 181. *See also* chronometers; jewelers; watchcases; watchmakers; watch pocket; watch register; *and names of individual watch manufacturers and individual watches*
watchmakers, *16*, 30, *65*, 81–84, 111–12, 122, *125*. *See also* jewelers; watches
watch movements. *See* watches: movements for

watch pocket, 68, 213n21
watch register, 64, 211n4
Waterbury Watch Company, 73, 76, 146
Western Union Telegraph Company, 100–102, *101*, 104–5, 108, 141
Westminster chimes, 25, 196n6
Willard, Simon, 39, 49–50
Willard Brothers, 33, 34
work time, 46
world's fairs. *See individual fairs*

zones, time. *See* standard time: zones

CPSIA information can be obtained
at www.ICGtesting.com
Printed in the USA
LVHW082154140220
647003LV00012B/354